Science Pop Art
CELLS
A terminology coloring book

Illustrated and written

by Mizzz Foster

Copyright © 2024 Melanie A. Foster
All rights reserved.

This book, or parts thereof may not
be reproduced in any form without
permission from the publisher; exceptions
are made for brief excerpts used
in published reviews.

Published by Mizzz Foster
Eastvale, CA US
Email: MizzzFoster@gmail.com

ISBN: 978-1-7377913-1-7

Printed in the United States of America
Design by M. A. Foster
Cover Design by M. A. Foster
Illustrations © 2024 M. A. Foster

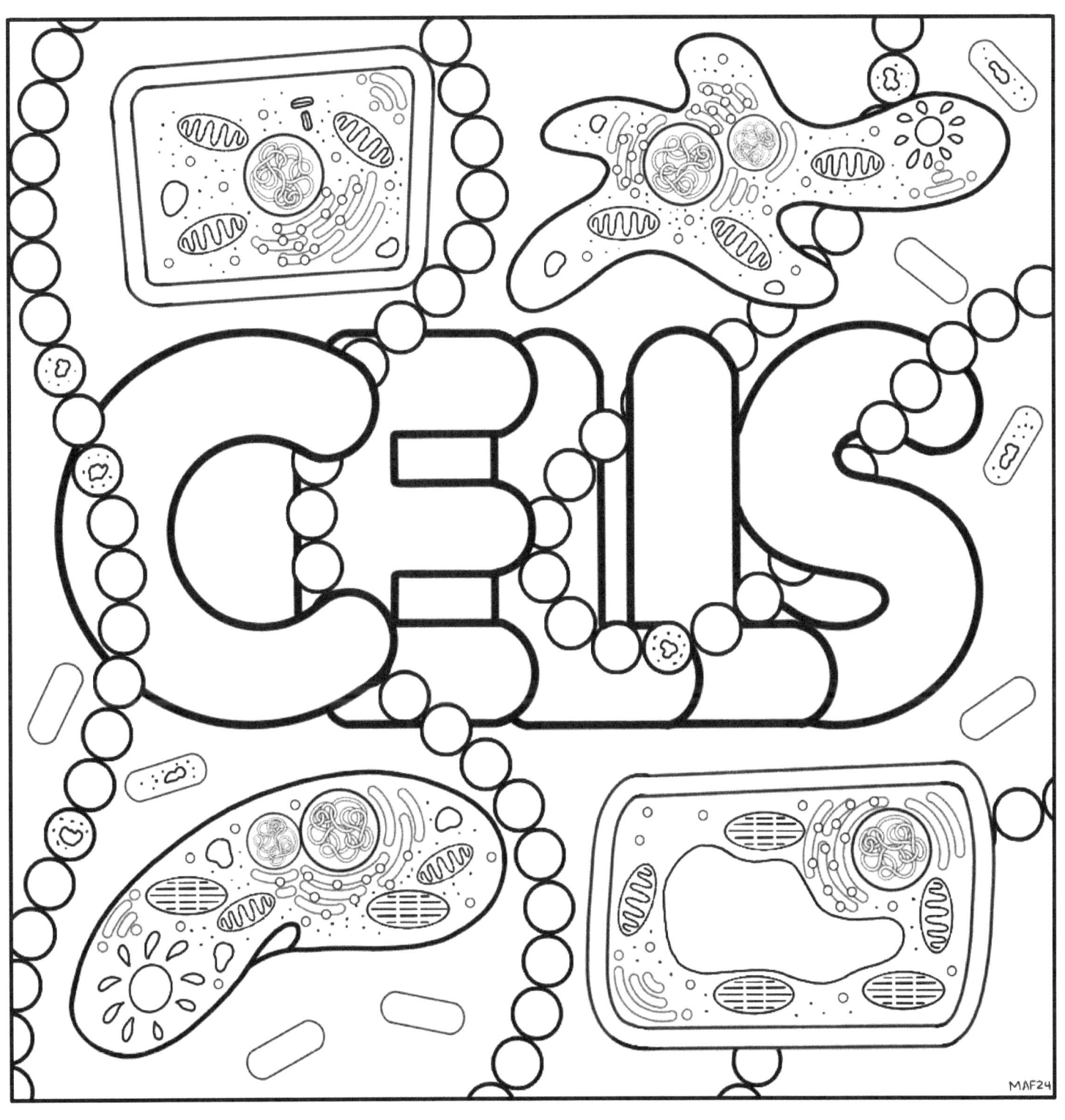

Definition: _____
_____
_____
_____
_____

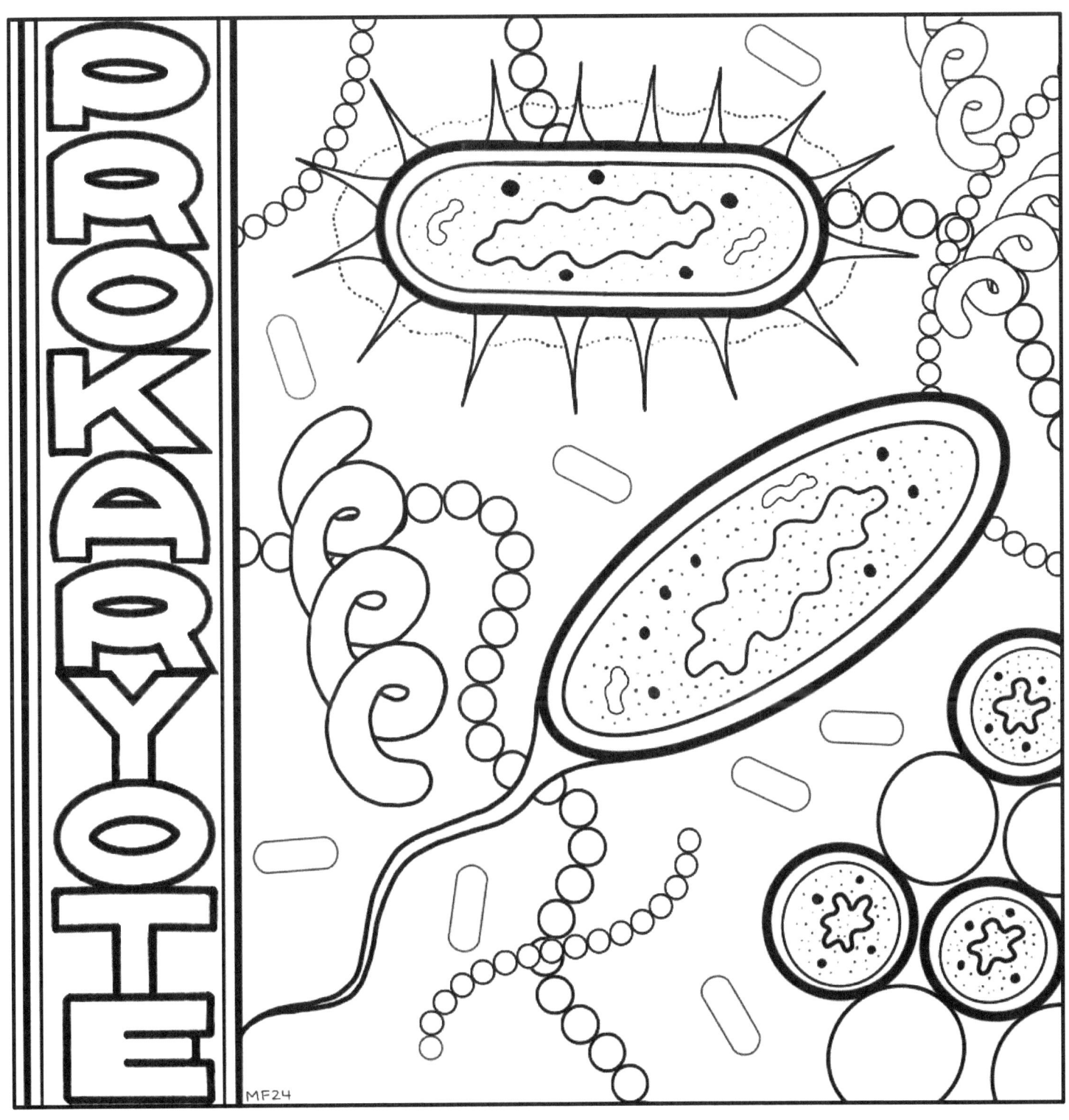

Definition: _____
_____
_____
_____
_____

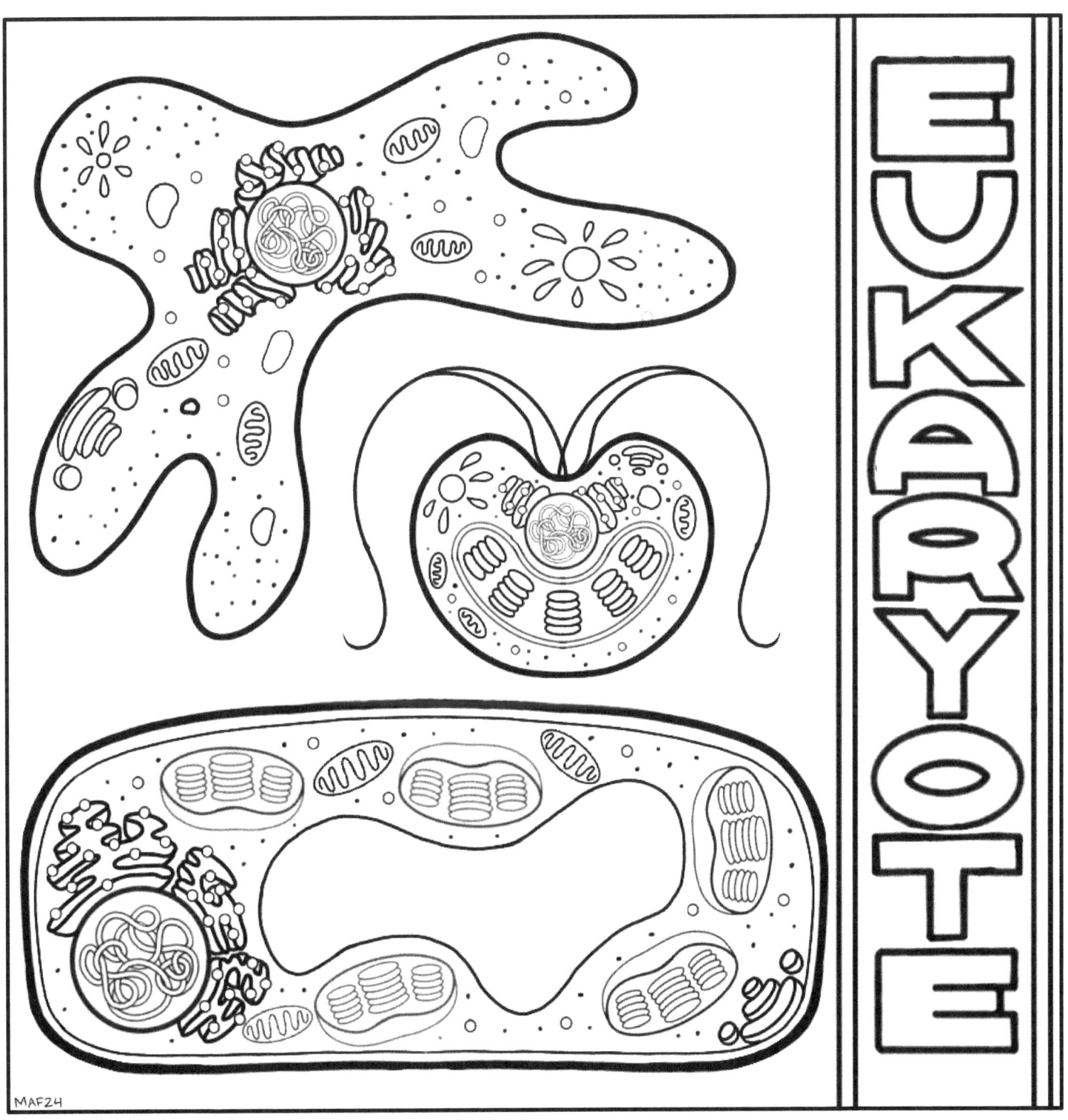

Definition: _____
_____
_____
_____
_____

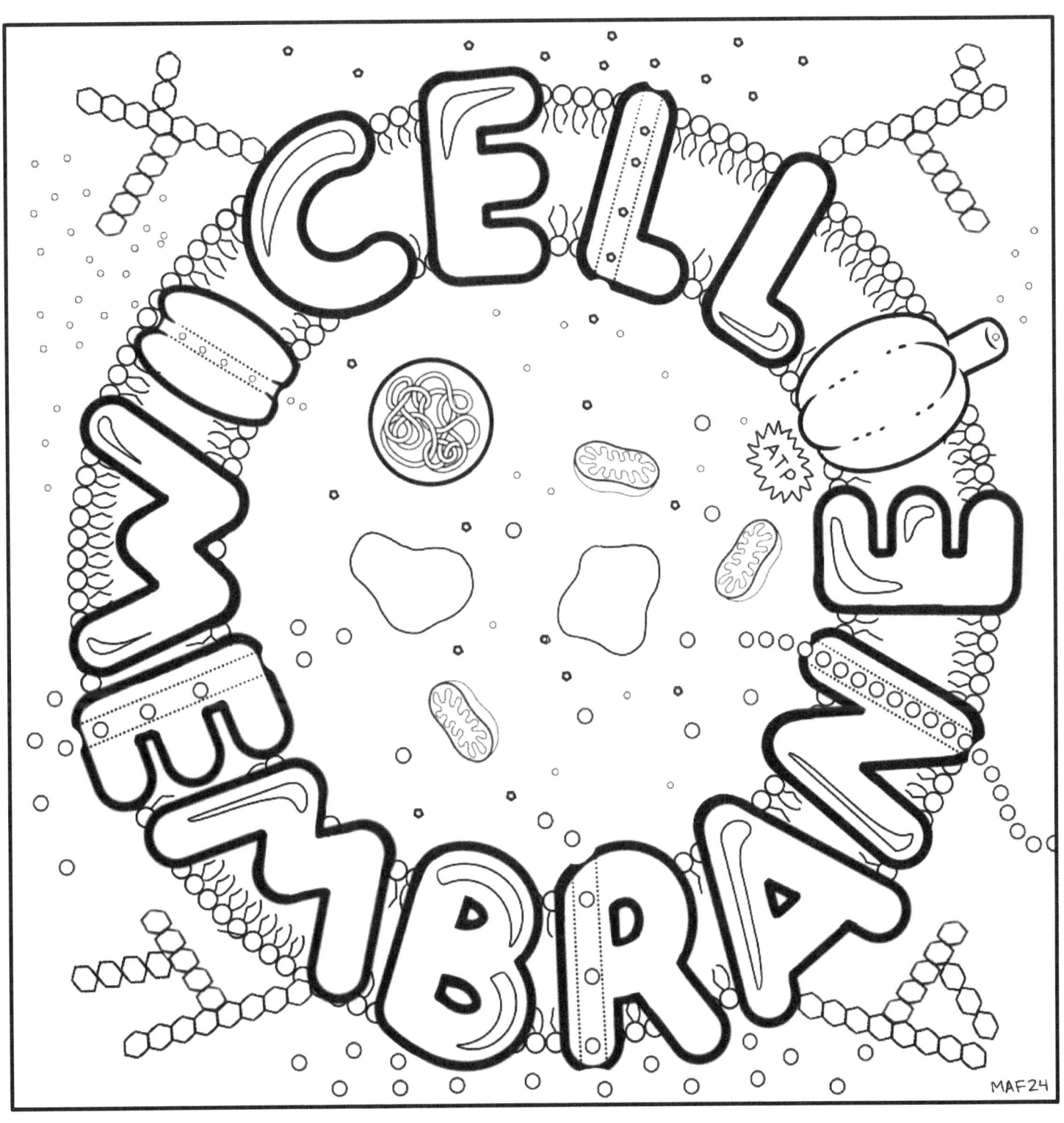

Definition: _____
_____
_____
_____
_____

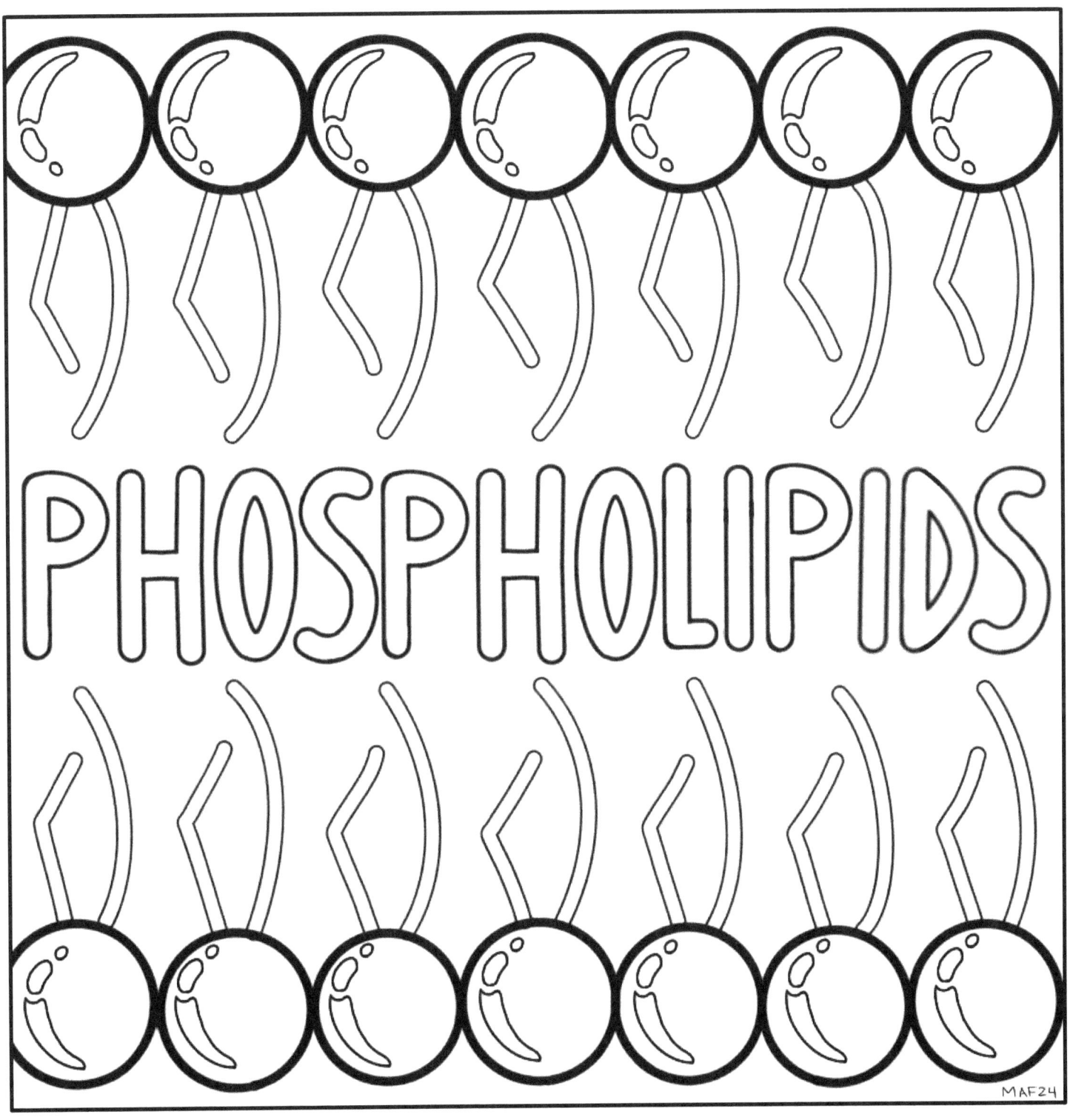

Definition: _____
_____
_____
_____
_____

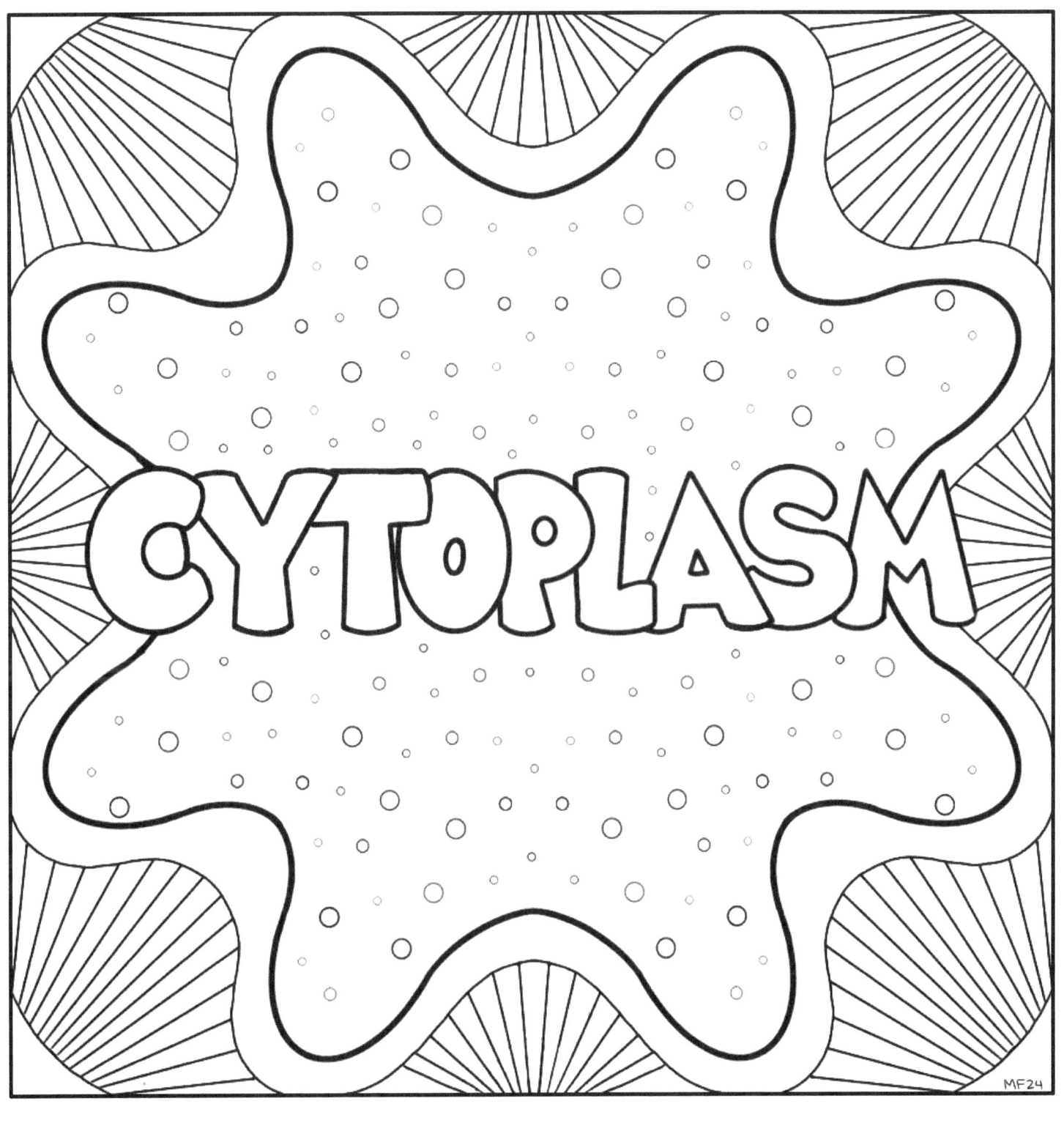

Definition: _____
_____
_____
_____
_____

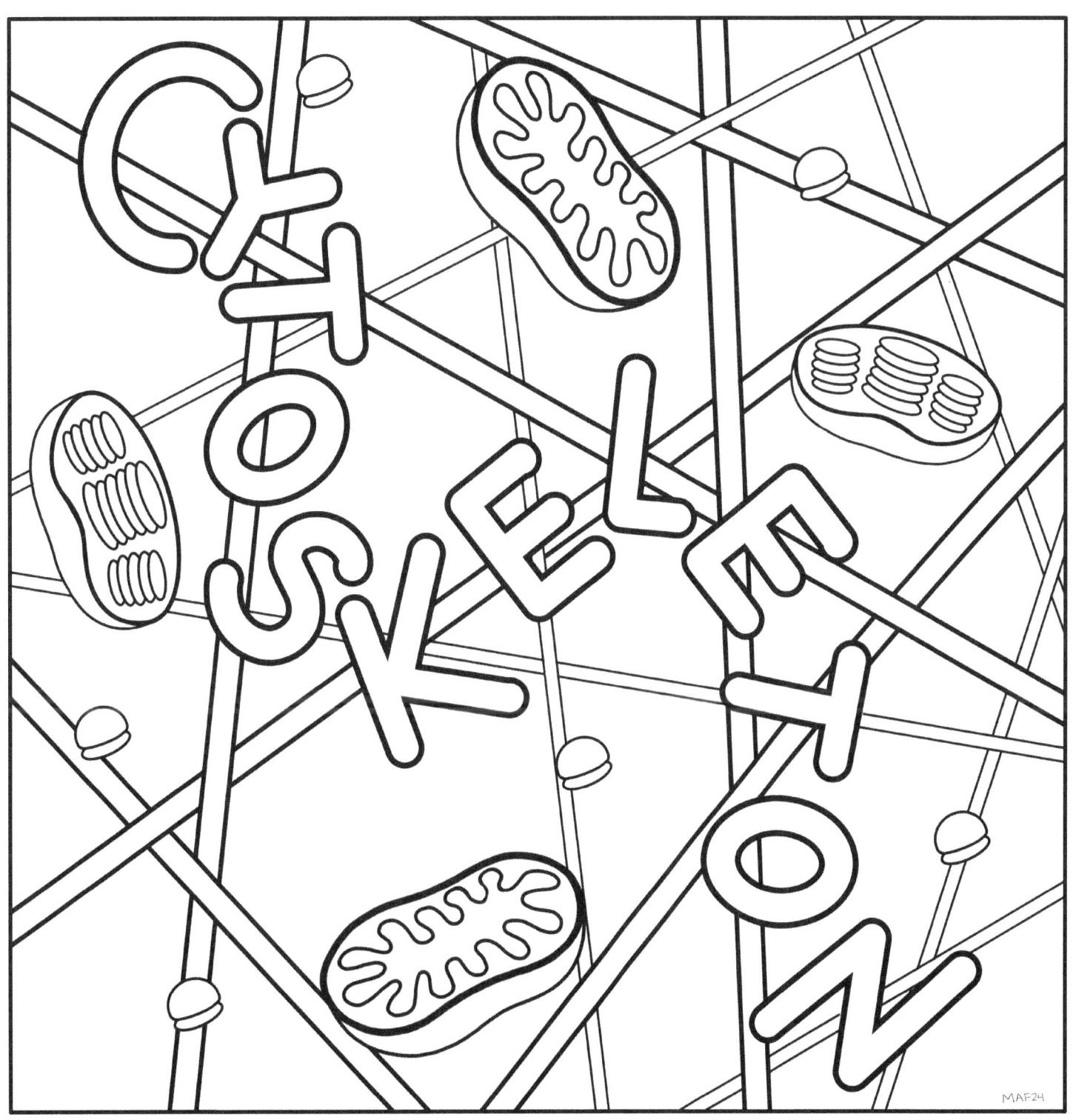

Definition: _____
_____
_____
_____
_____

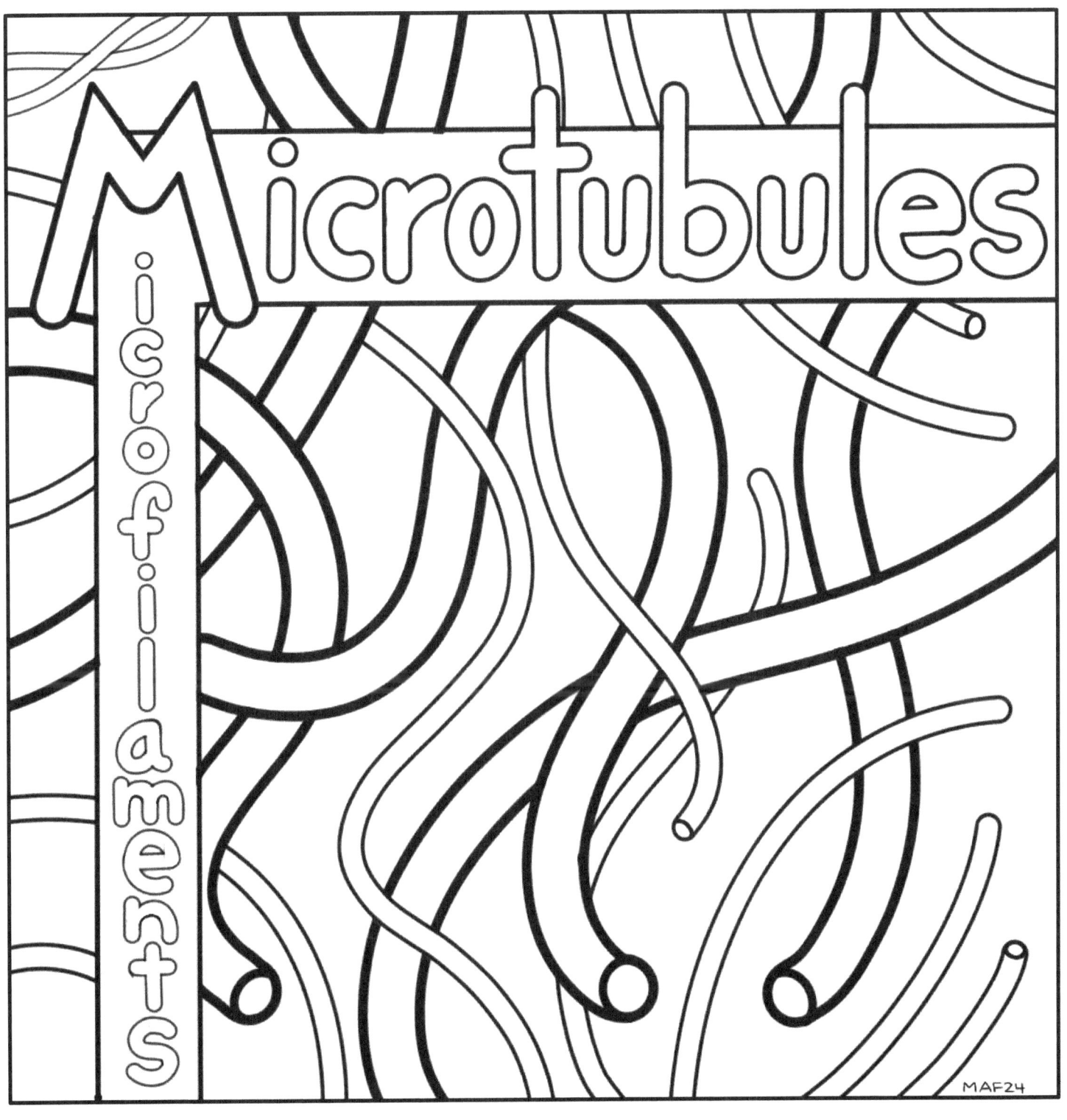

Definition: _____
_____
_____
_____
_____

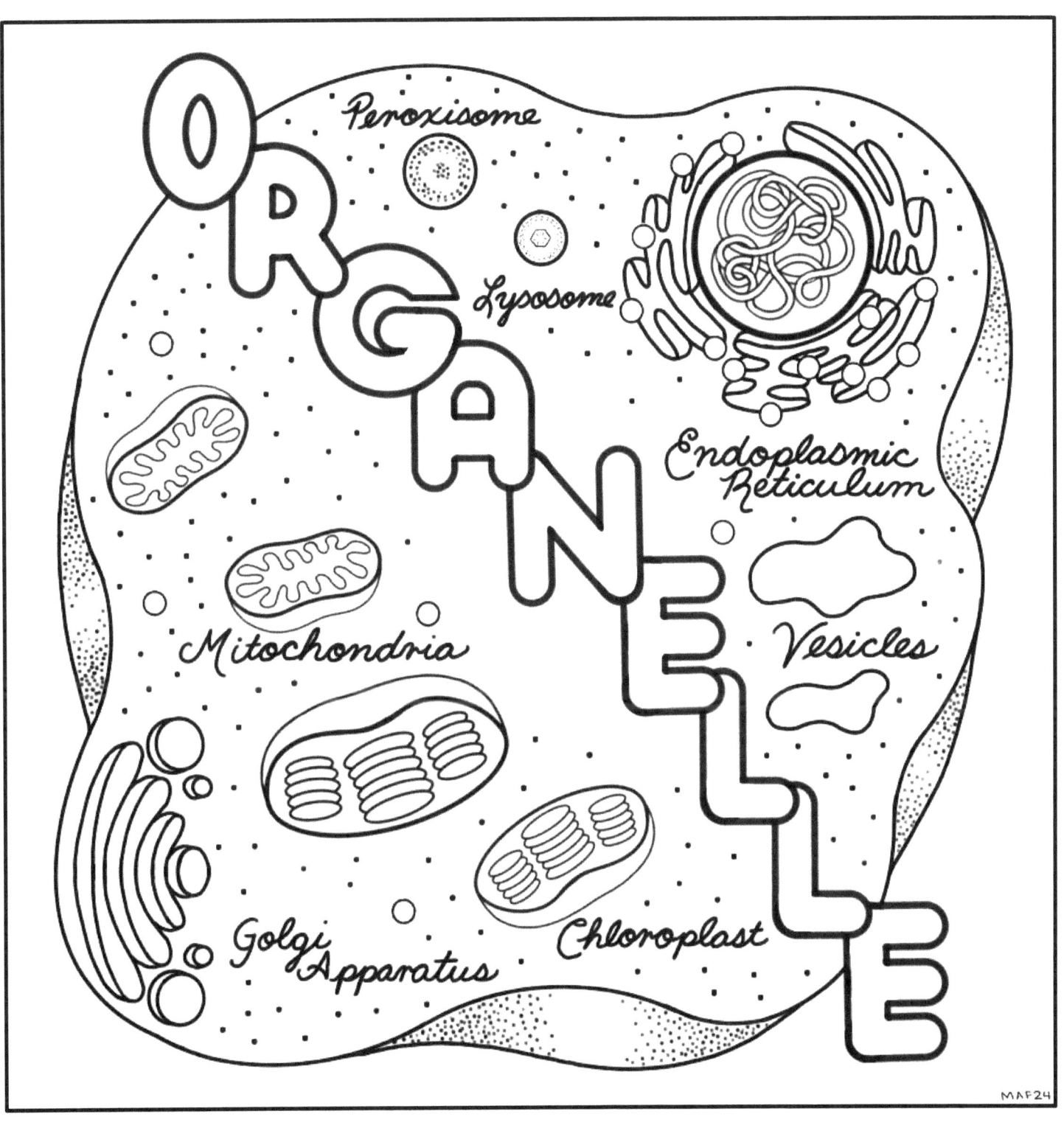

Definition: _____
_____
_____
_____
_____

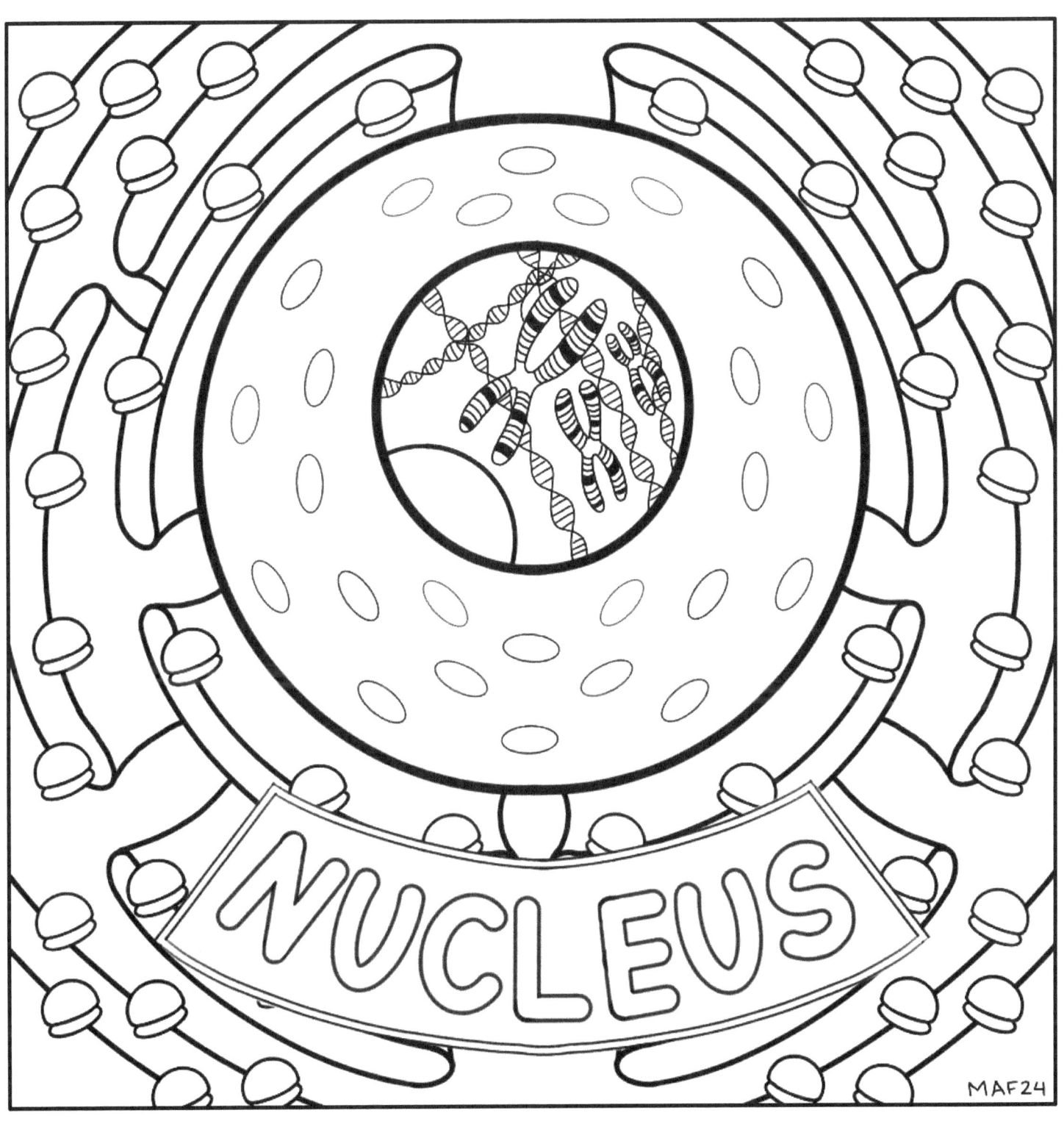

Definition: _____
_____
_____
_____
_____

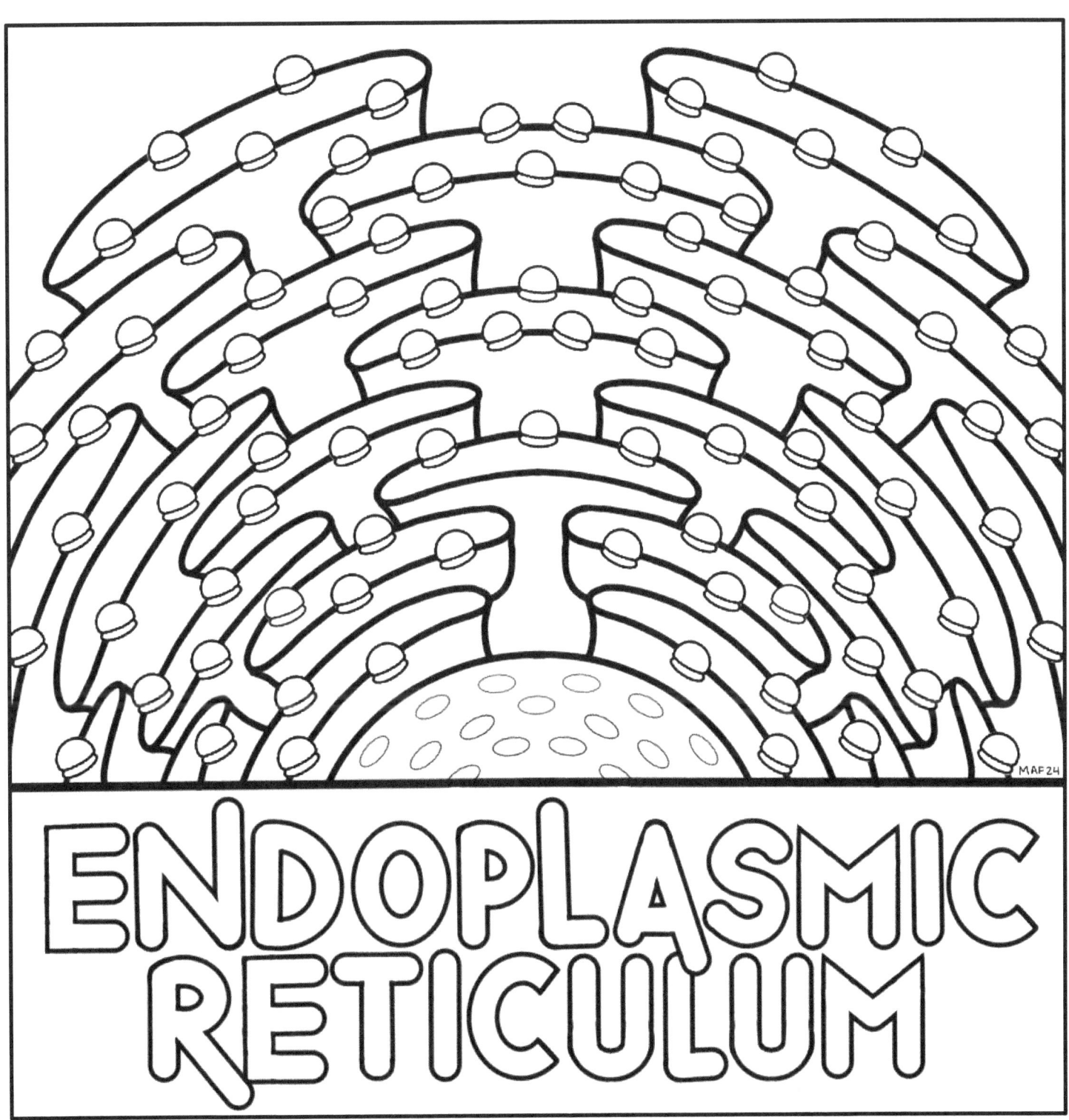

Definition: _____

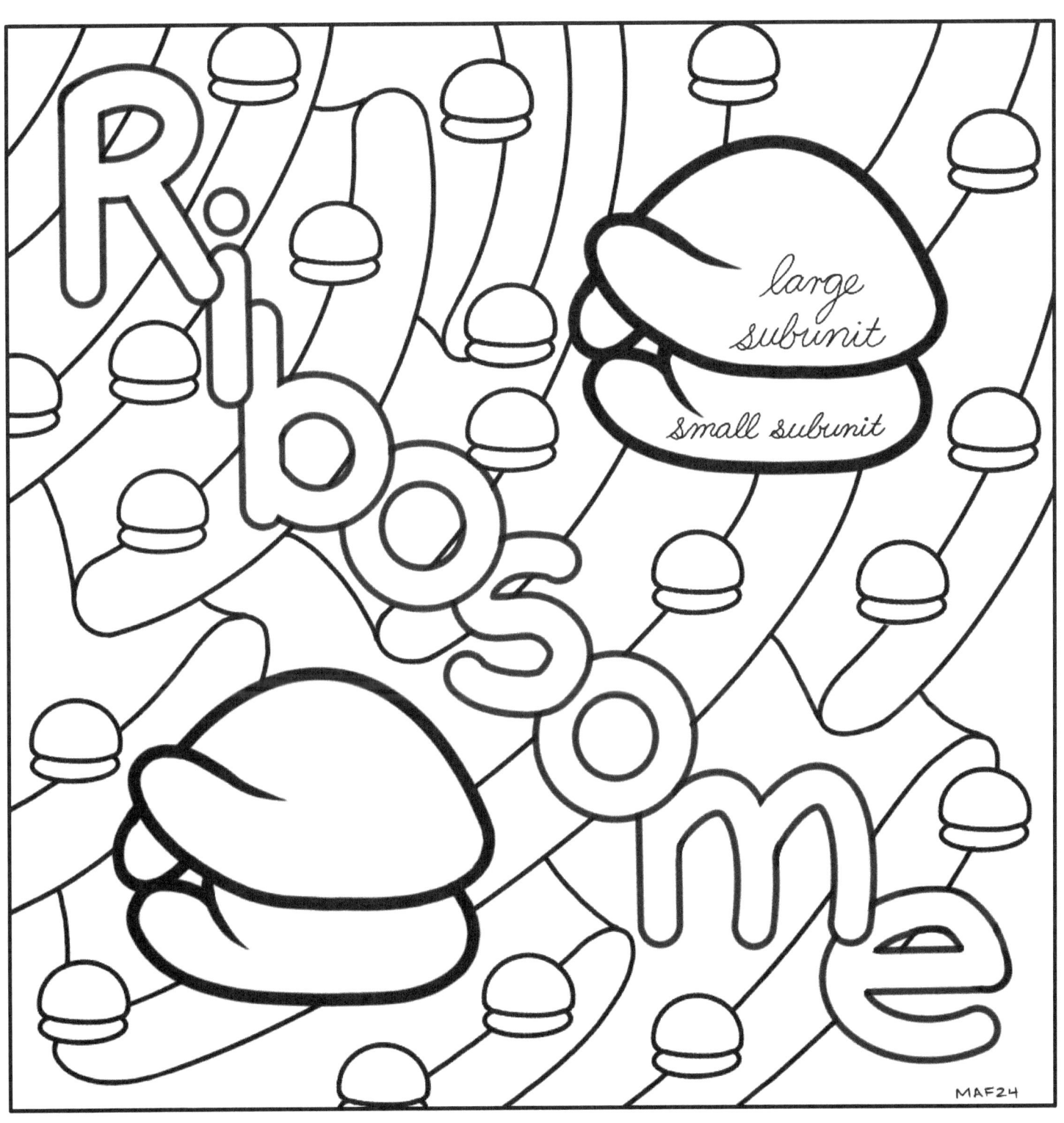

Definition: _____
_____
_____
_____
_____

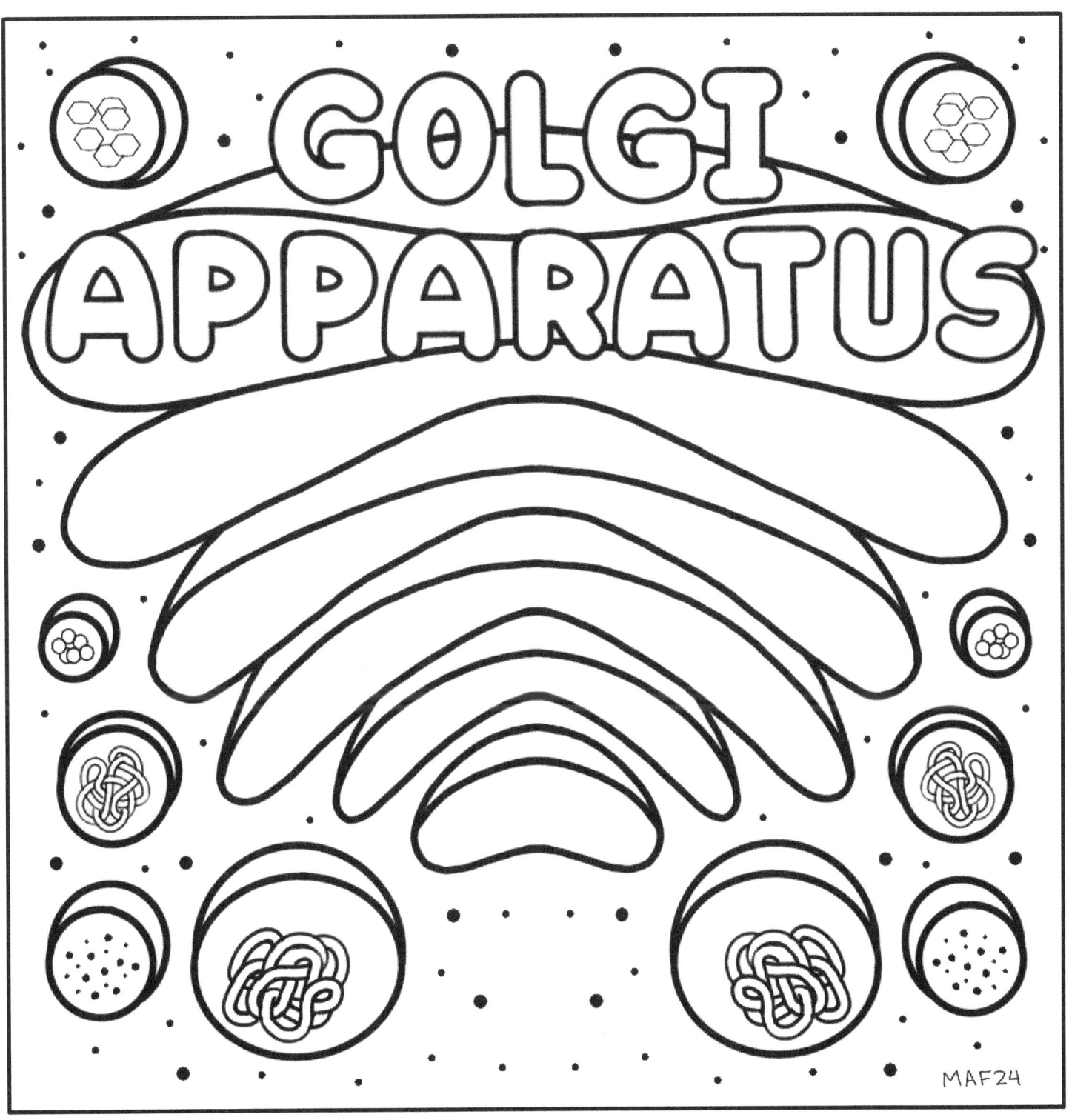

Definition: _____
_____
_____
_____
_____

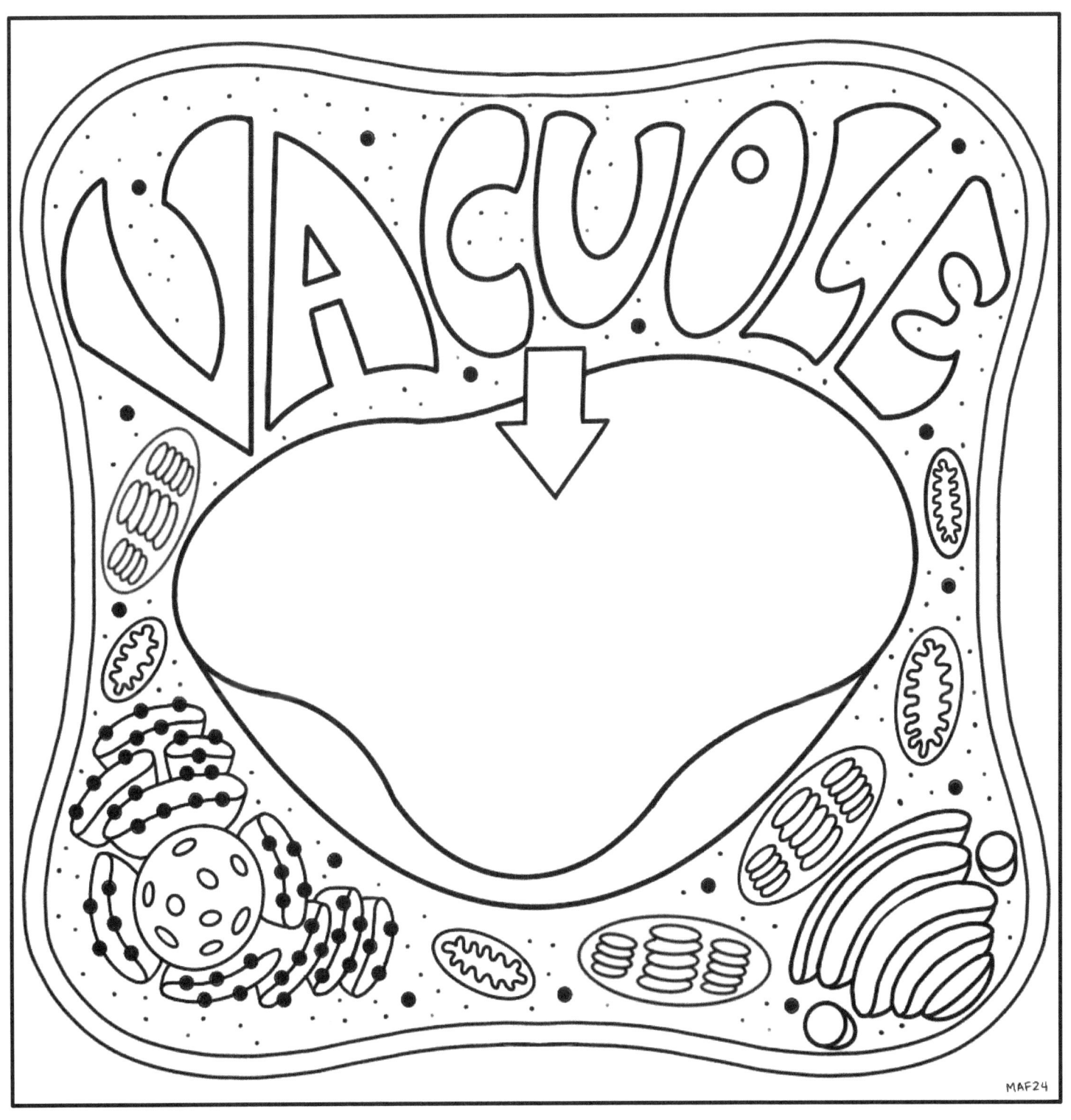

Definition: _____
_____
_____
_____
_____

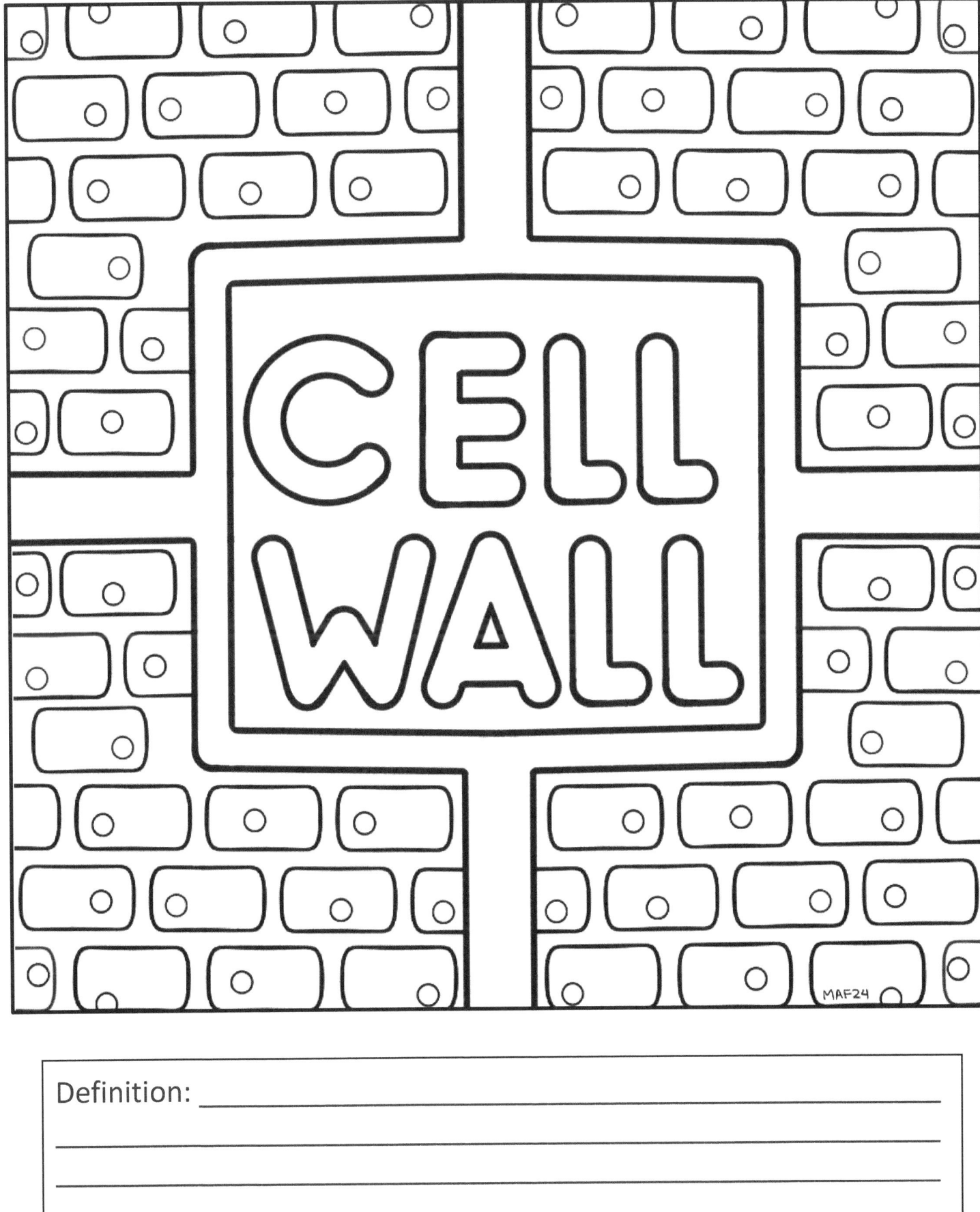

Definition: _____
_____
_____
_____
_____

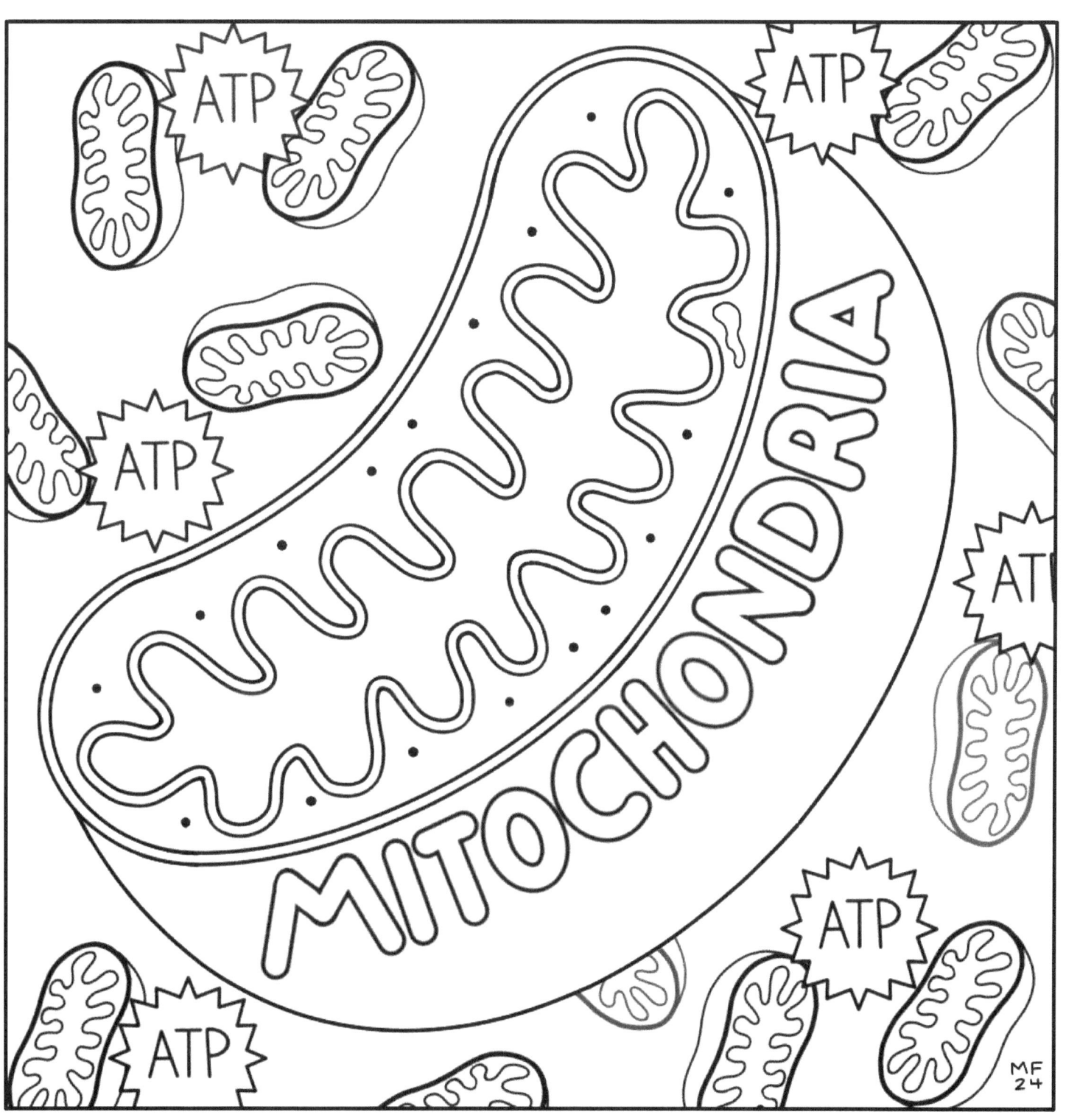

Definition: _____
_____
_____
_____
_____

Mizzz Foster © 2024

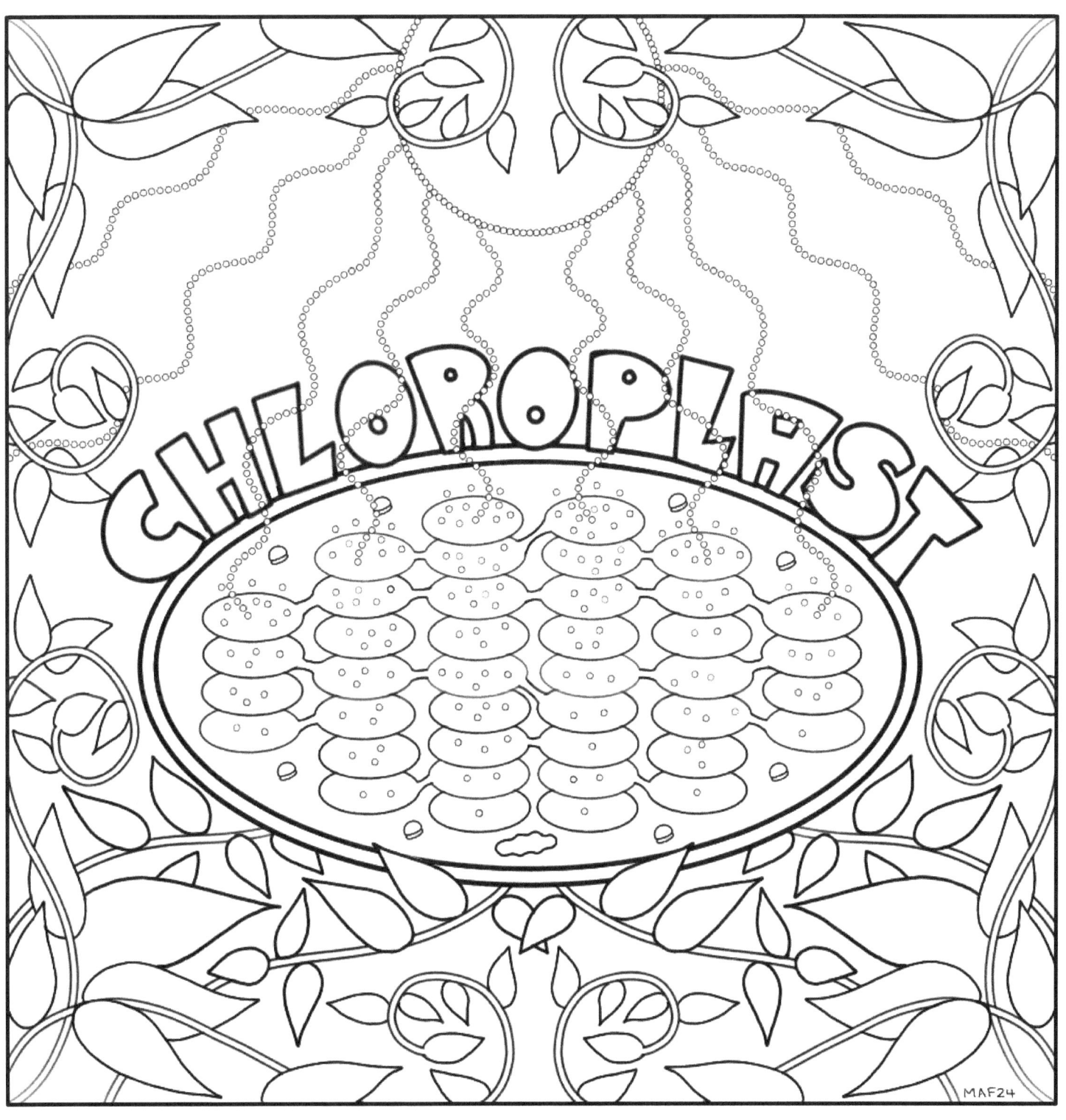

Definition: _____
_____
_____
_____
_____

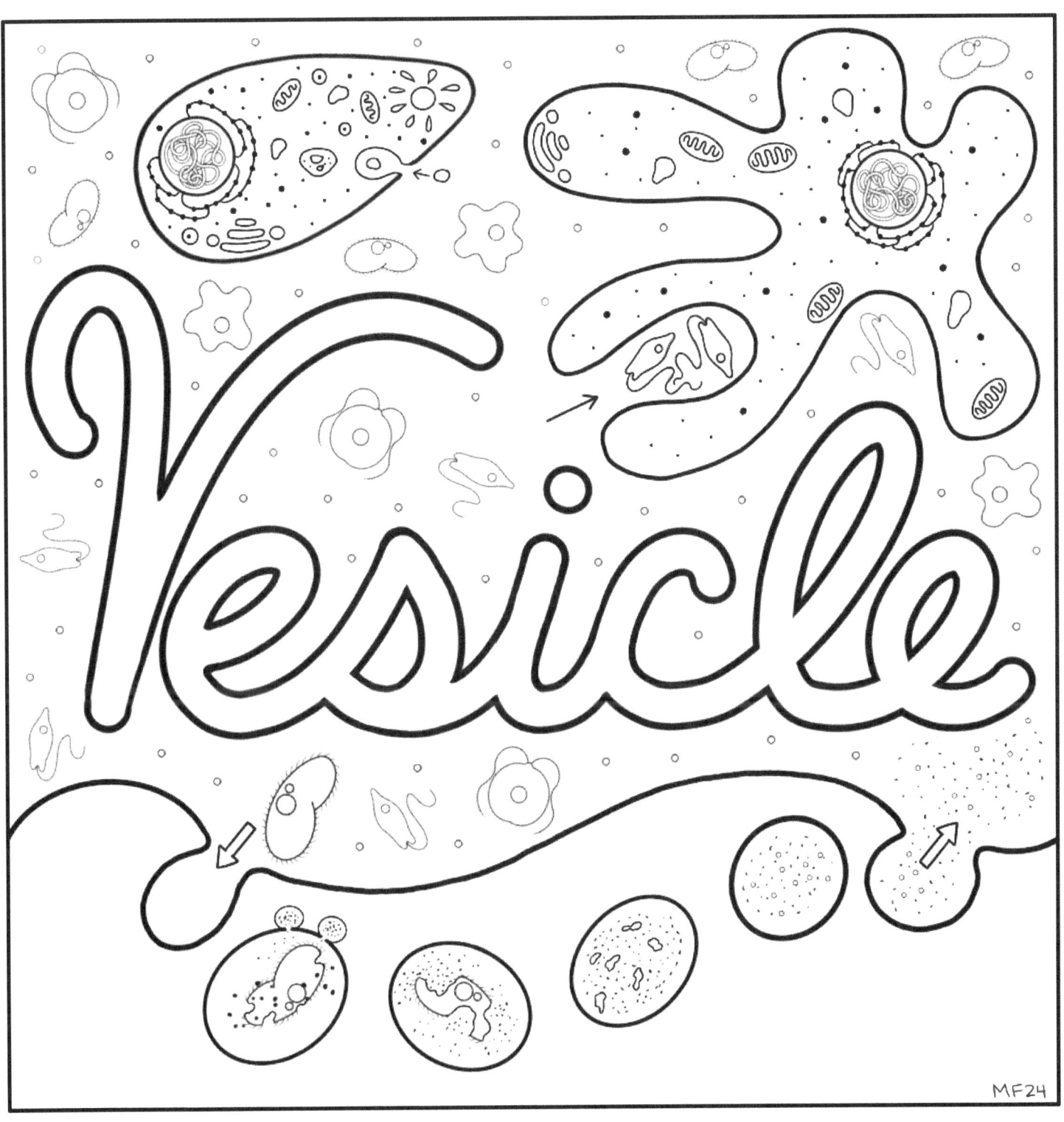

Definition: _____
_____
_____
_____
_____

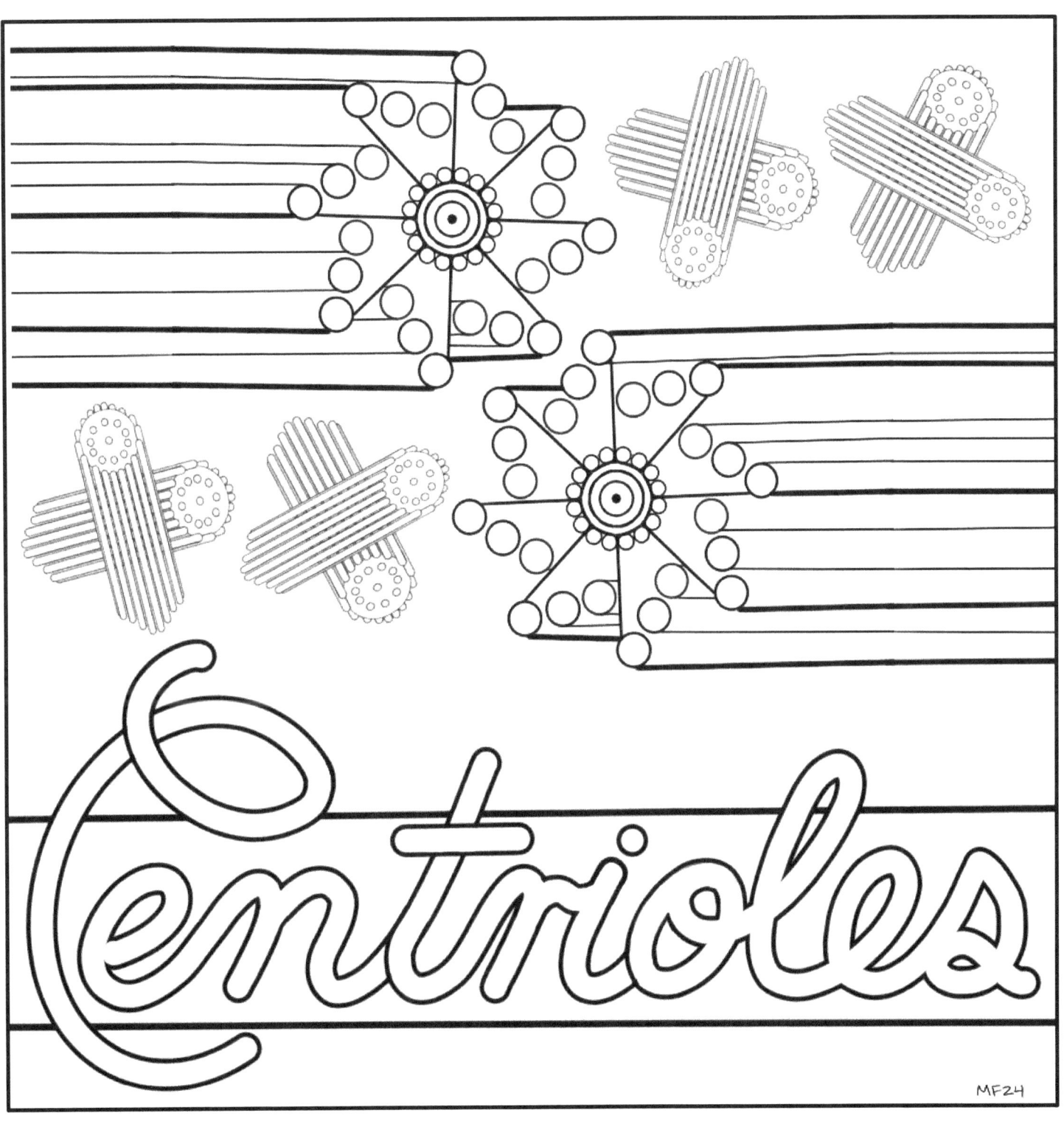

Definition: _____
_____
_____
_____
_____

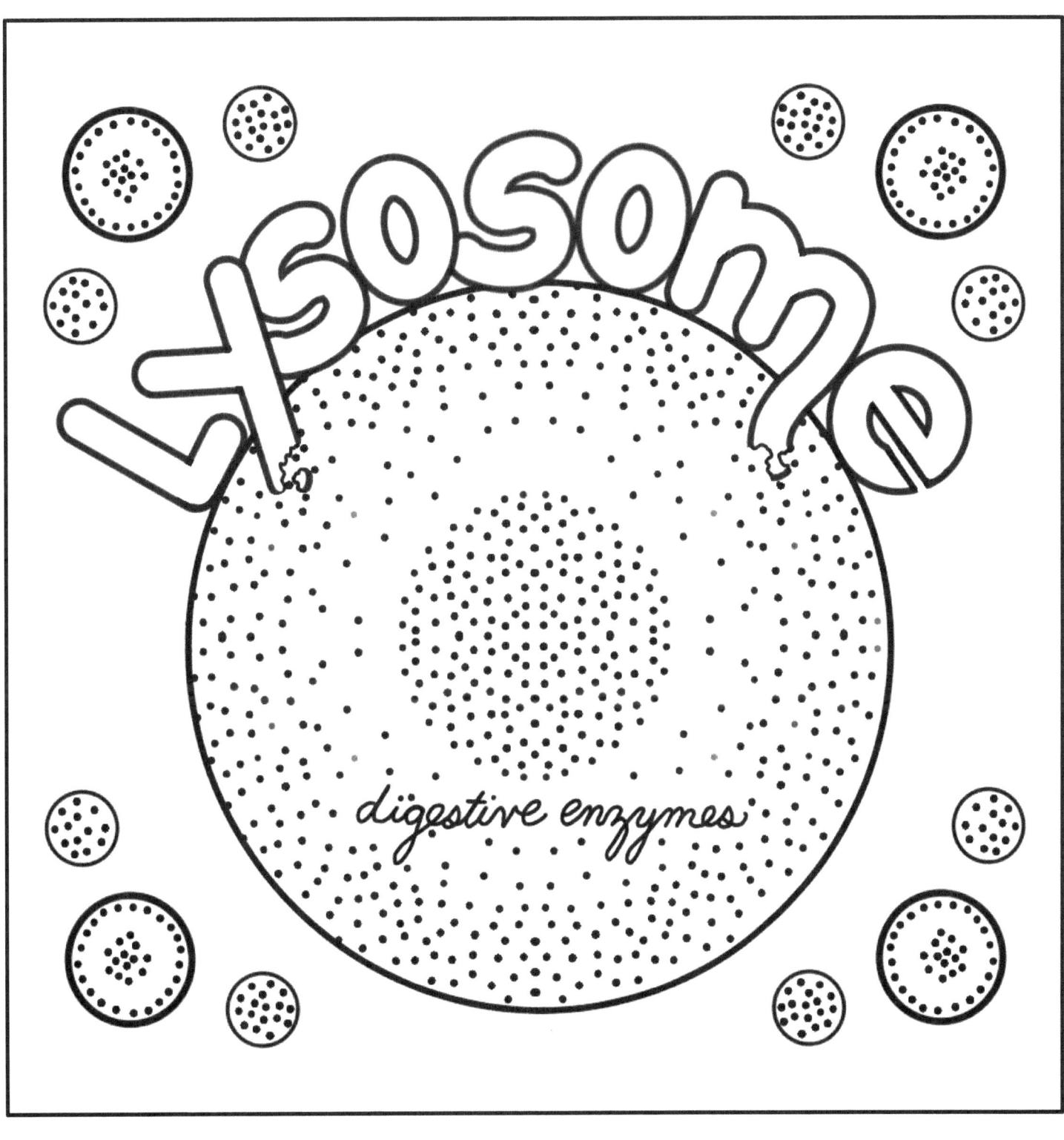

Definition: _____
_____
_____
_____
_____

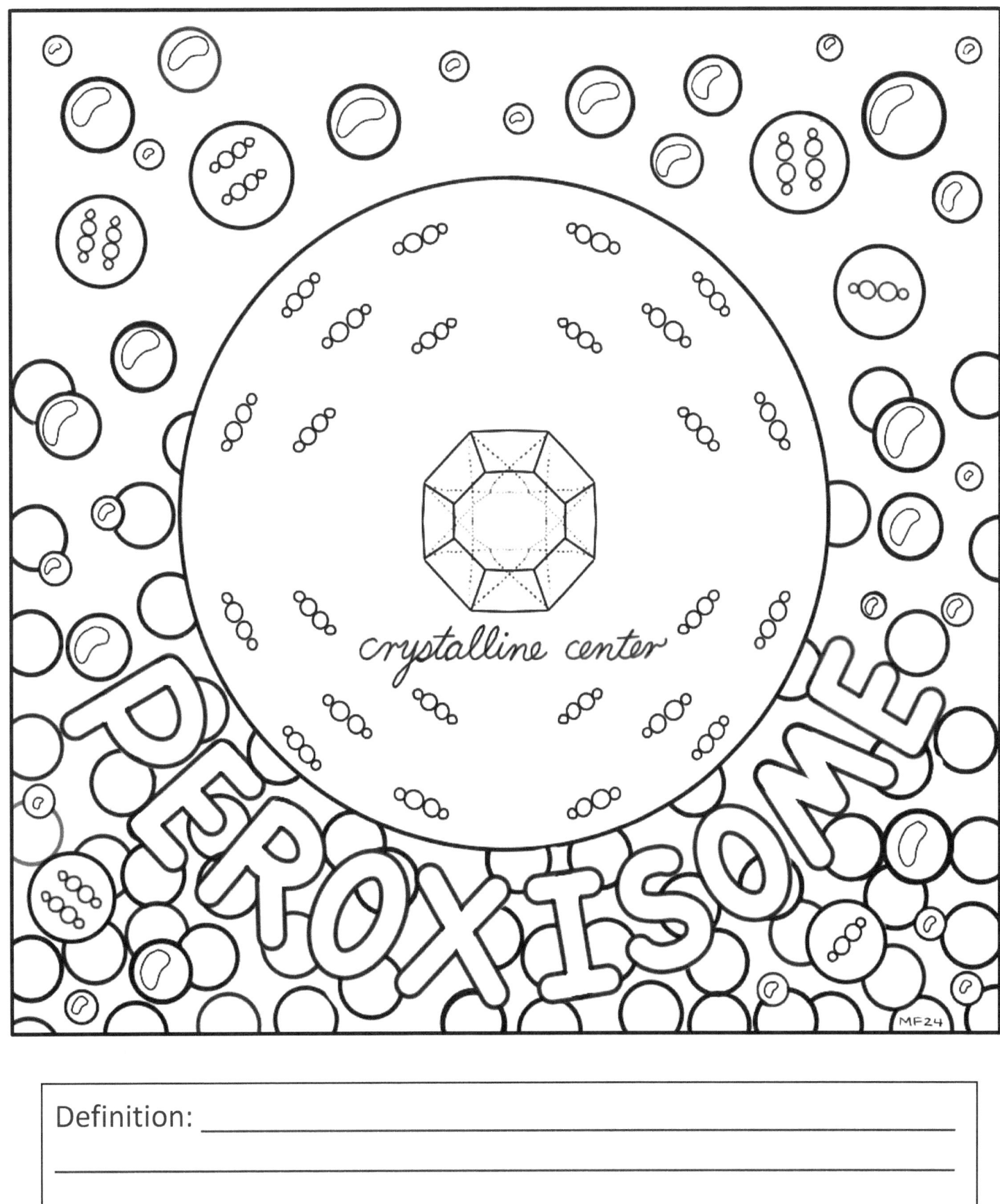

Definition: _____
_____
_____
_____
_____

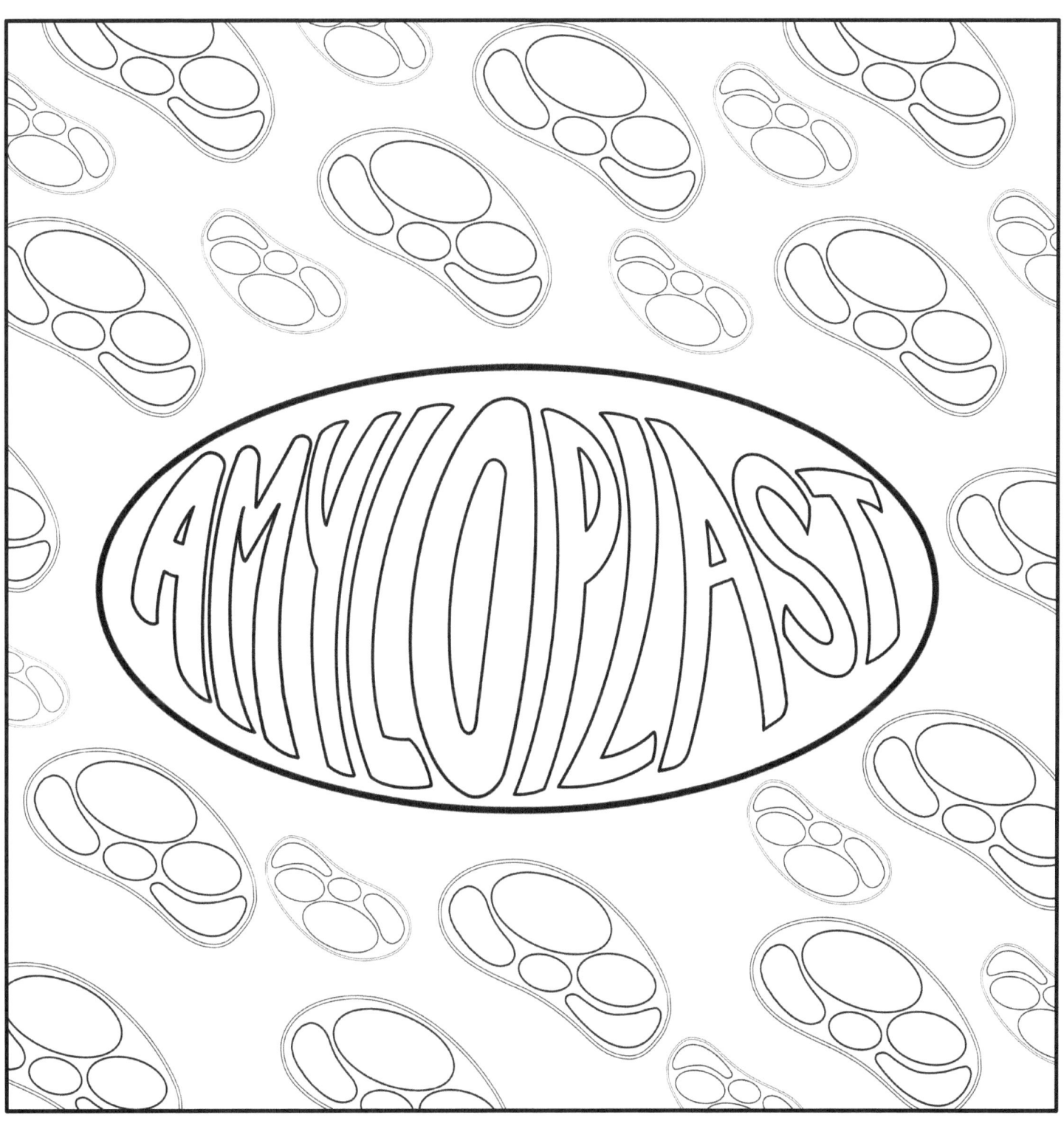

Definition: _____
_____
_____
_____
_____

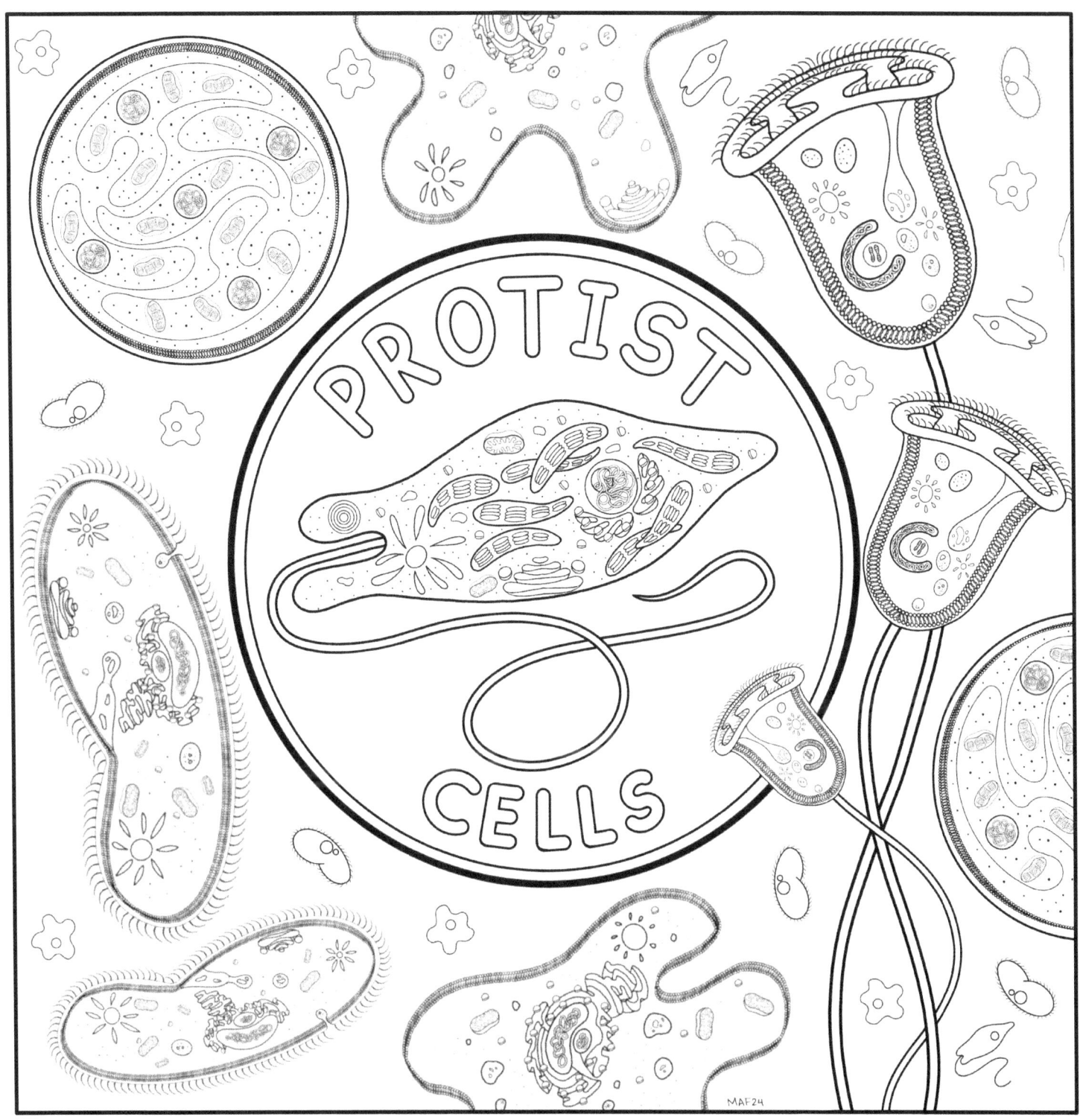

Definition: _____
_____
_____
_____
_____

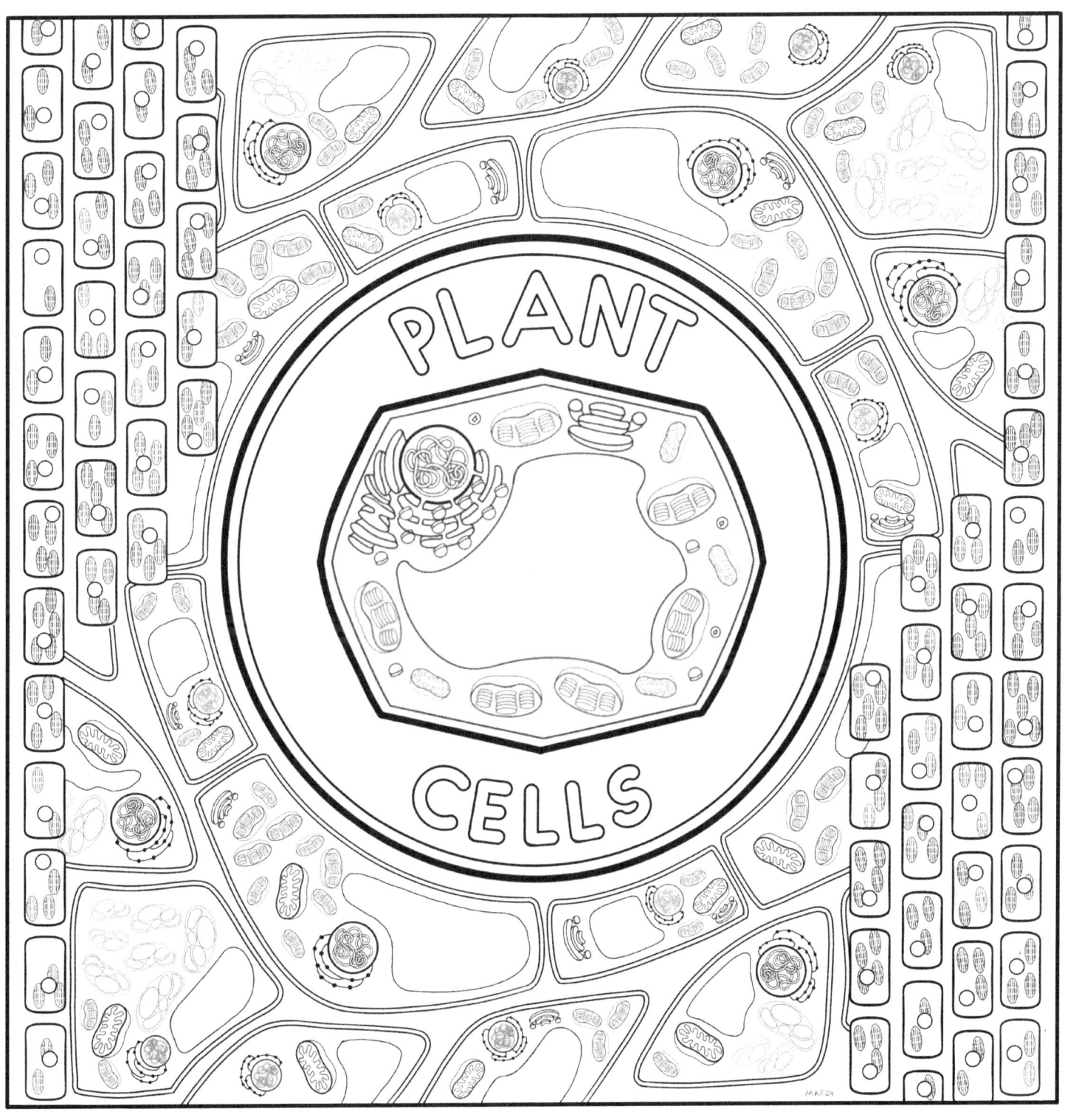

Definition: _____
_____
_____
_____
_____

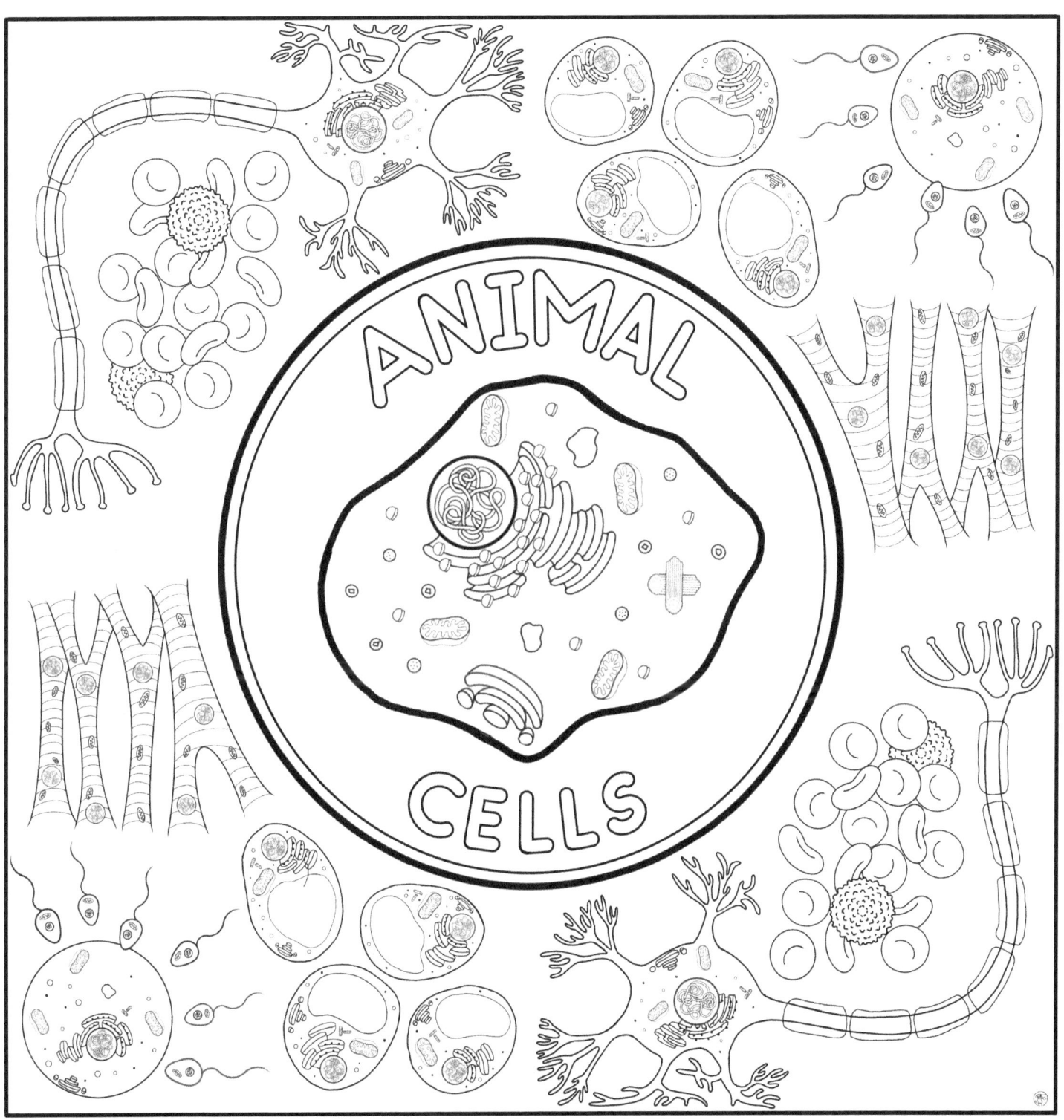

Definition: _____
_____
_____
_____
_____

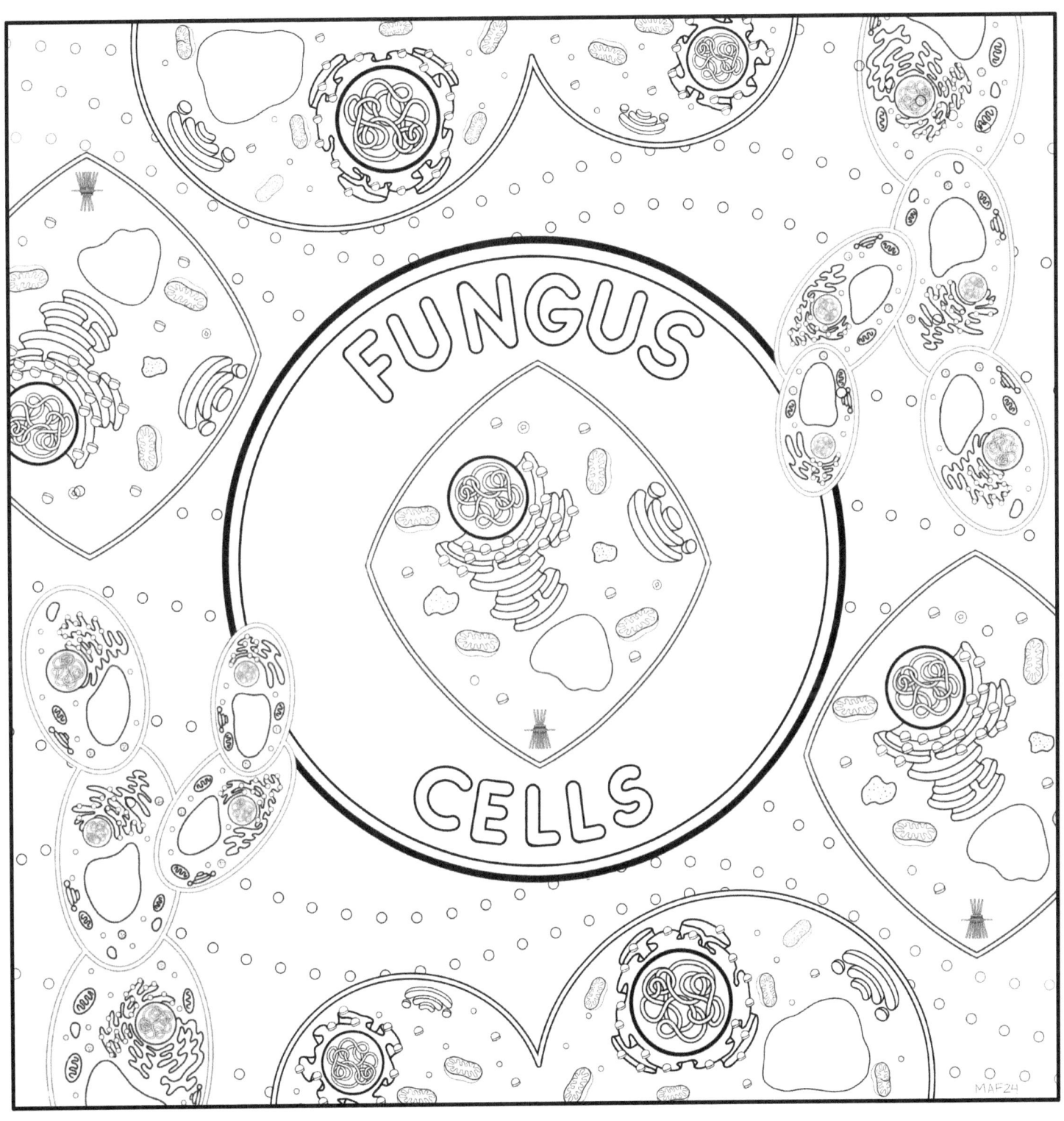

Definition: _____
_____
_____
_____
_____

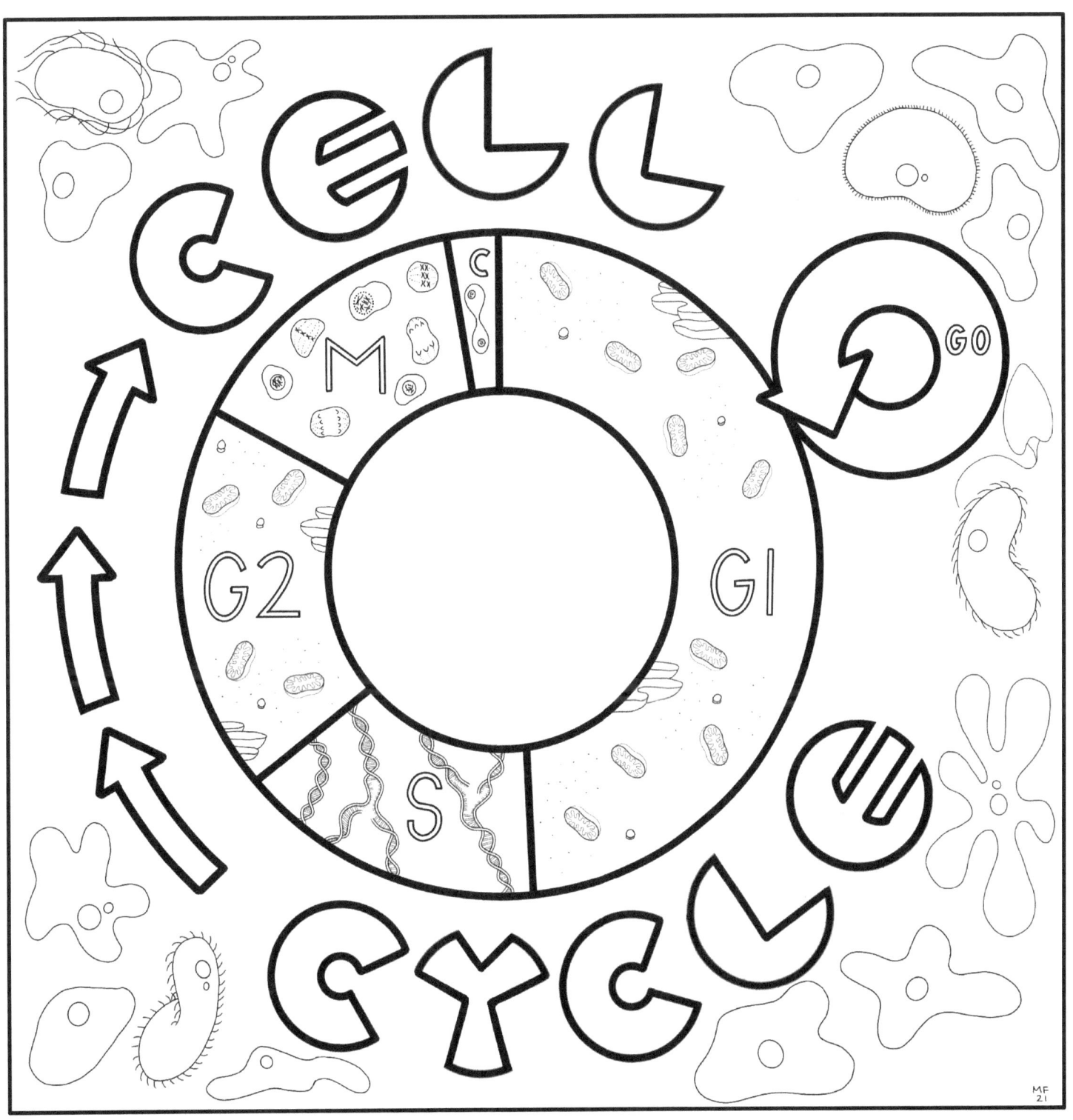

Definition: _____
_____
_____
_____
_____

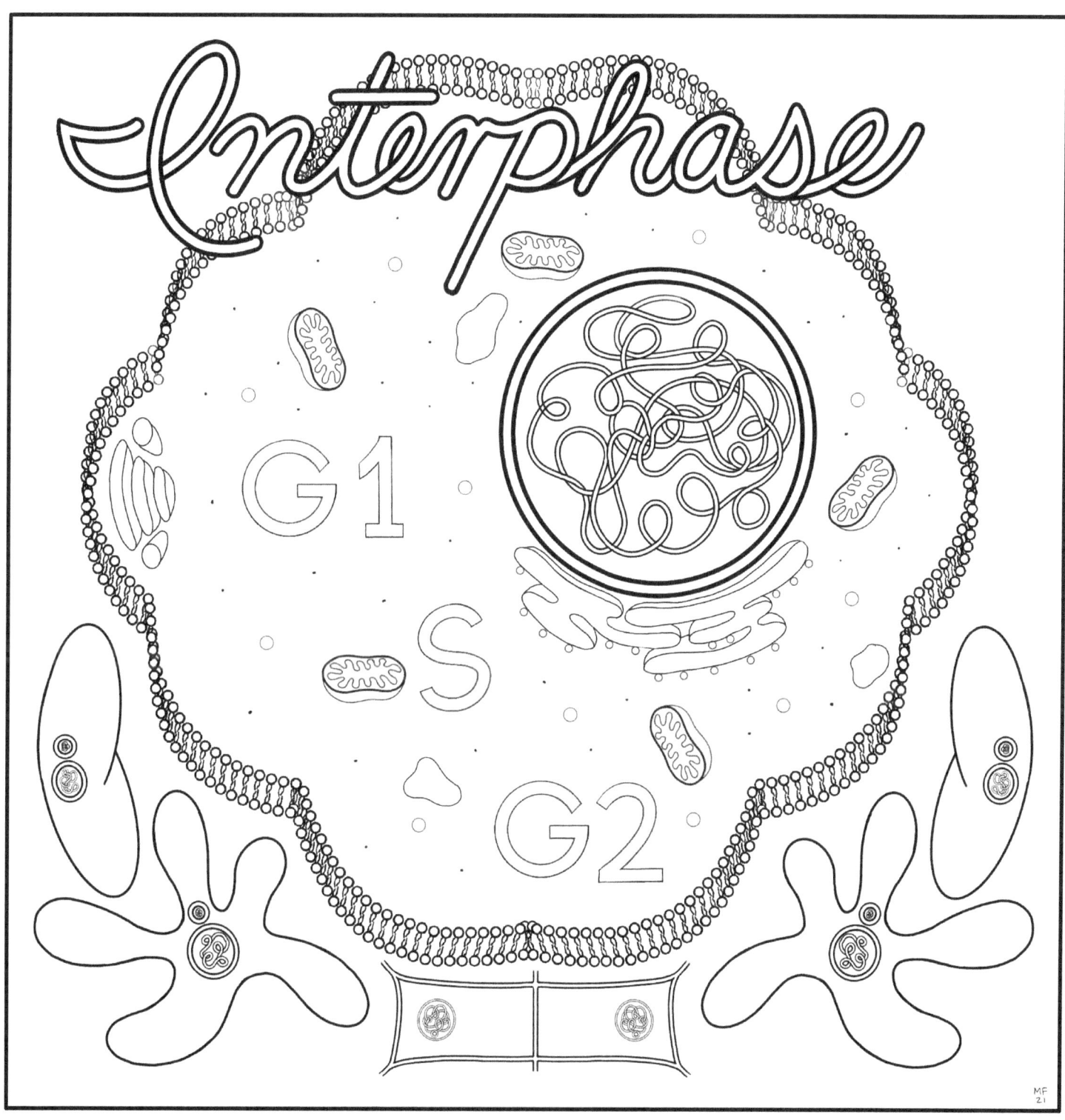

Definition: _____
_____
_____
_____
_____

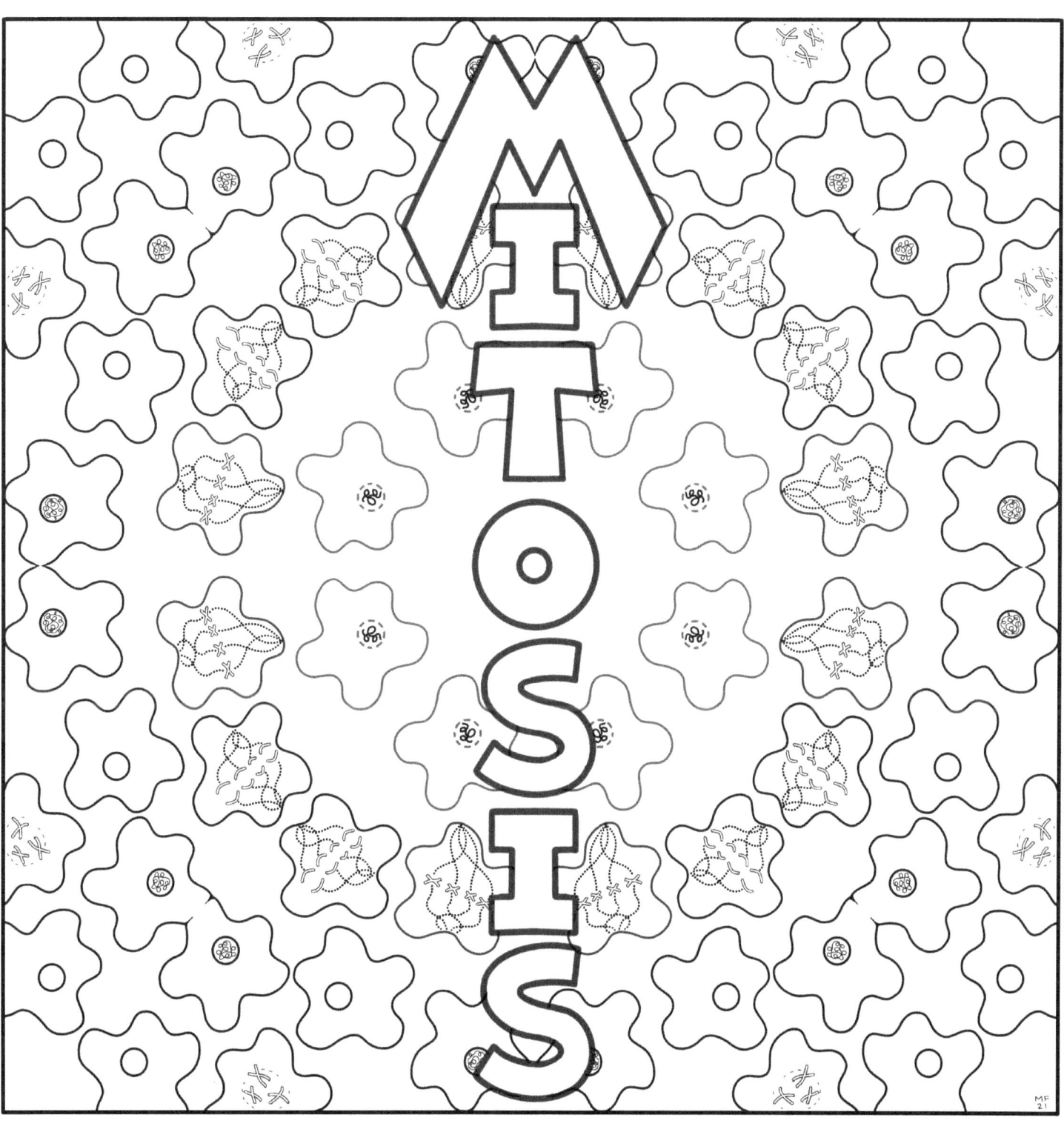

Definition: _____
_____
_____
_____
_____

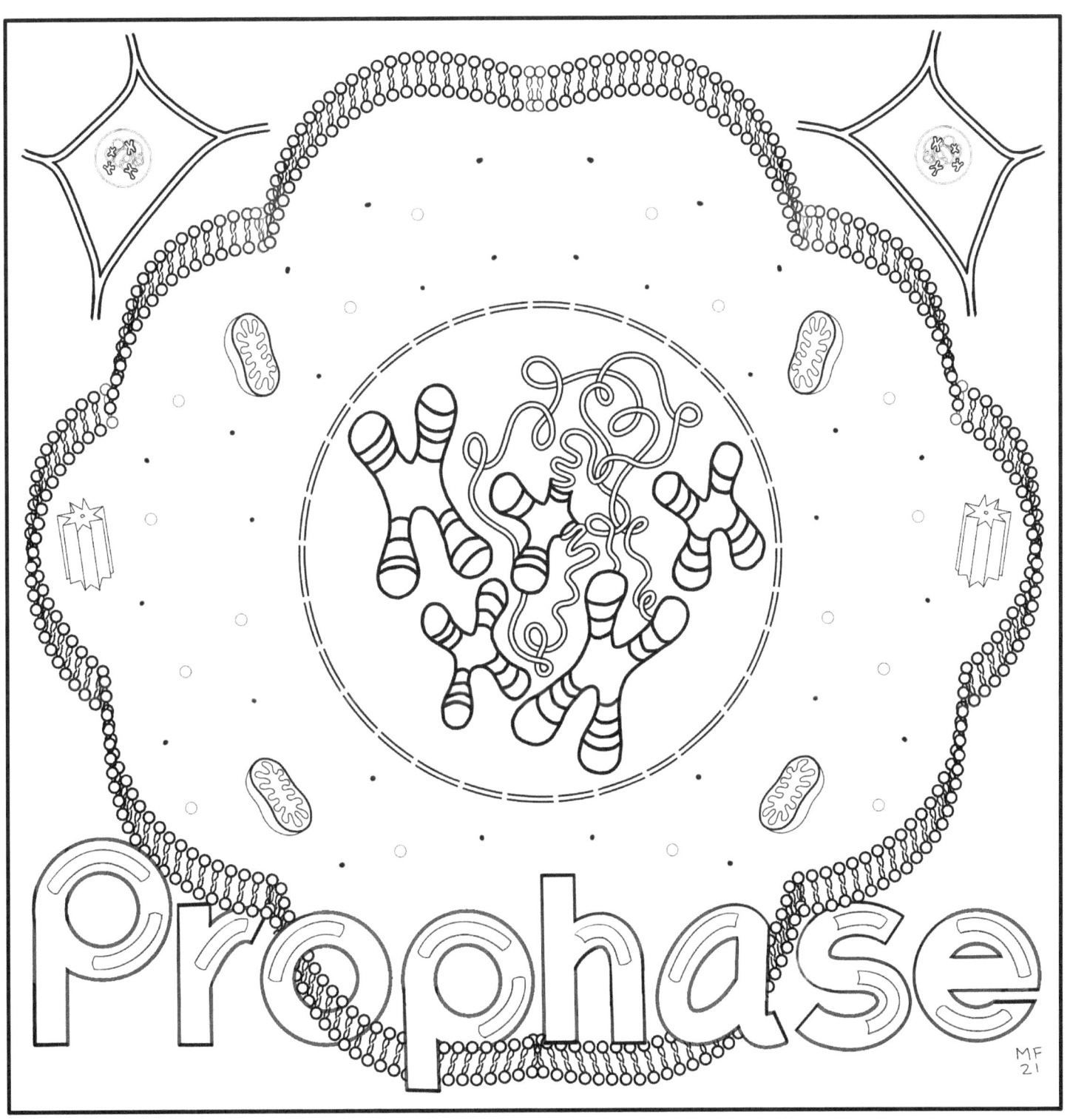

Definition: _____
_____
_____
_____
_____

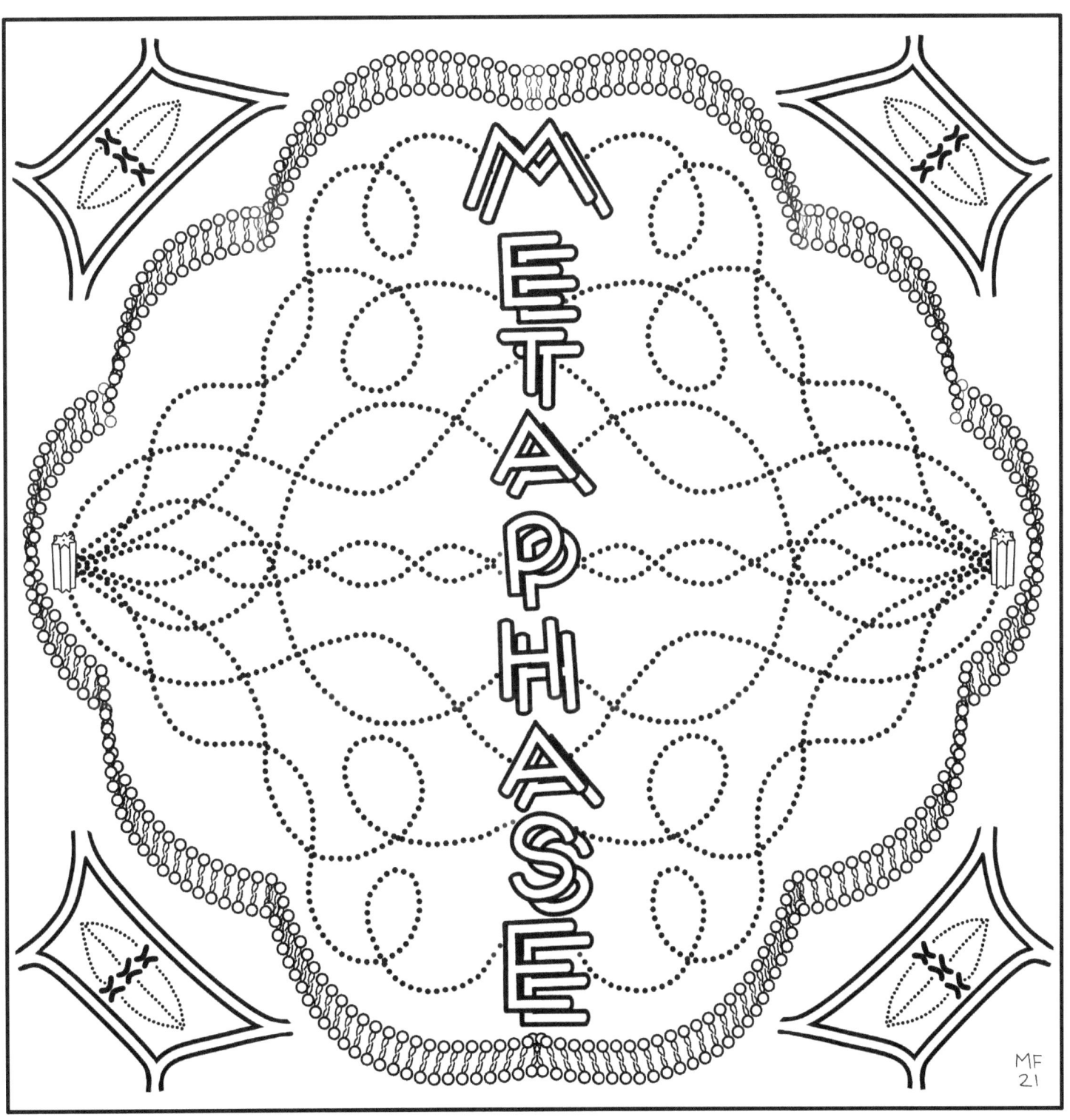

Definition: _____
_____
_____
_____
_____

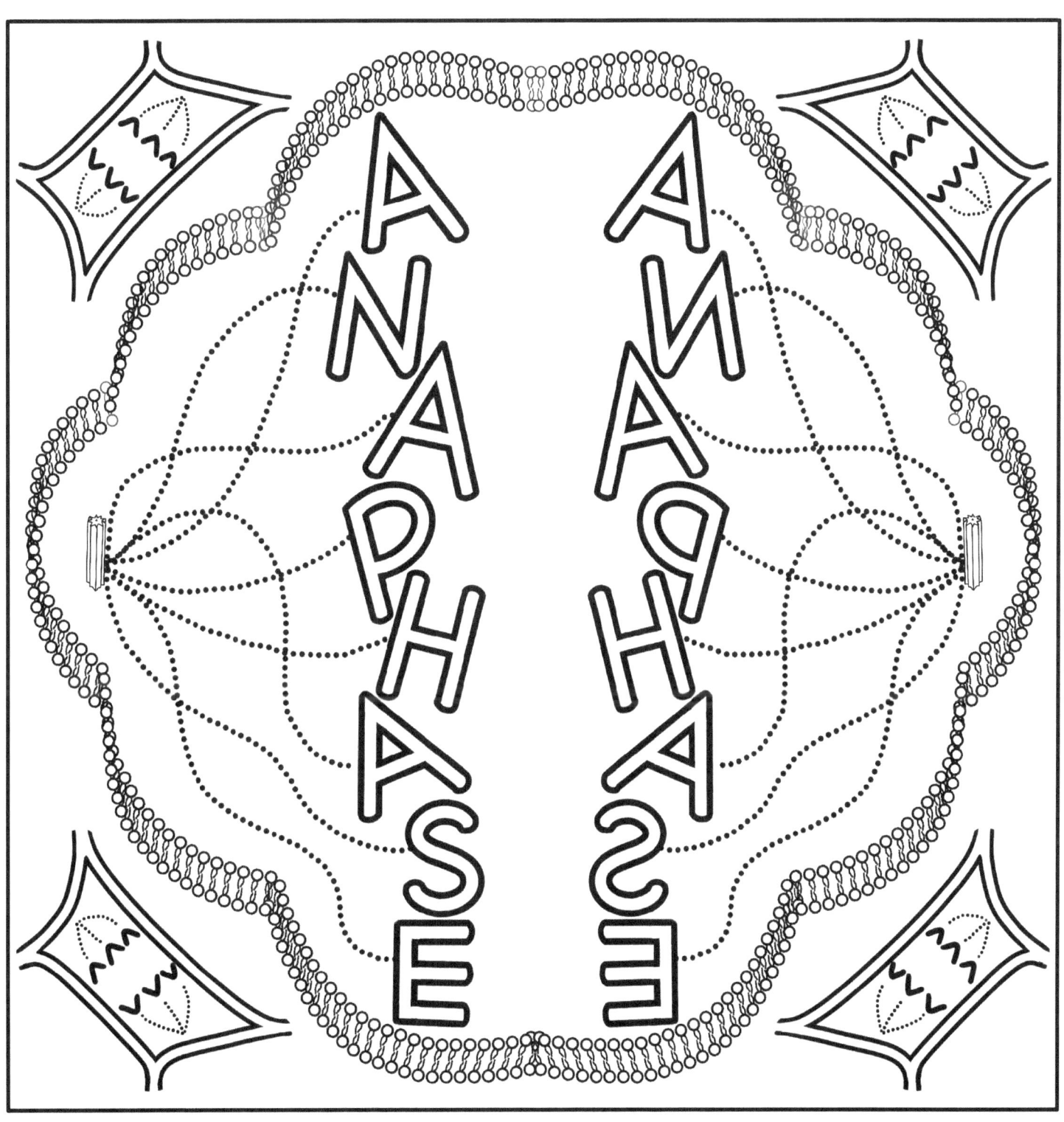

Definition: _____

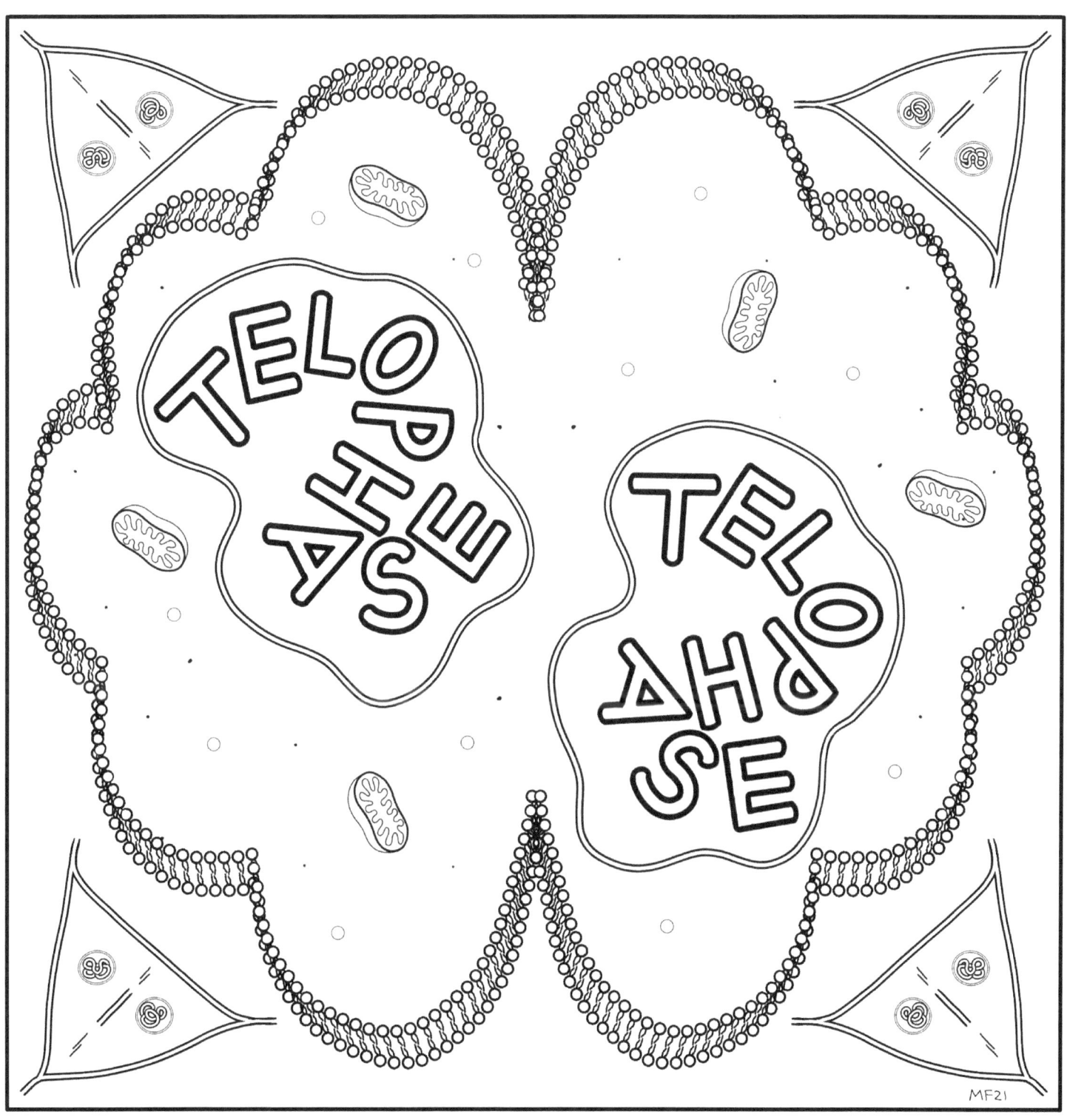

Definition: _____
_____
_____
_____
_____

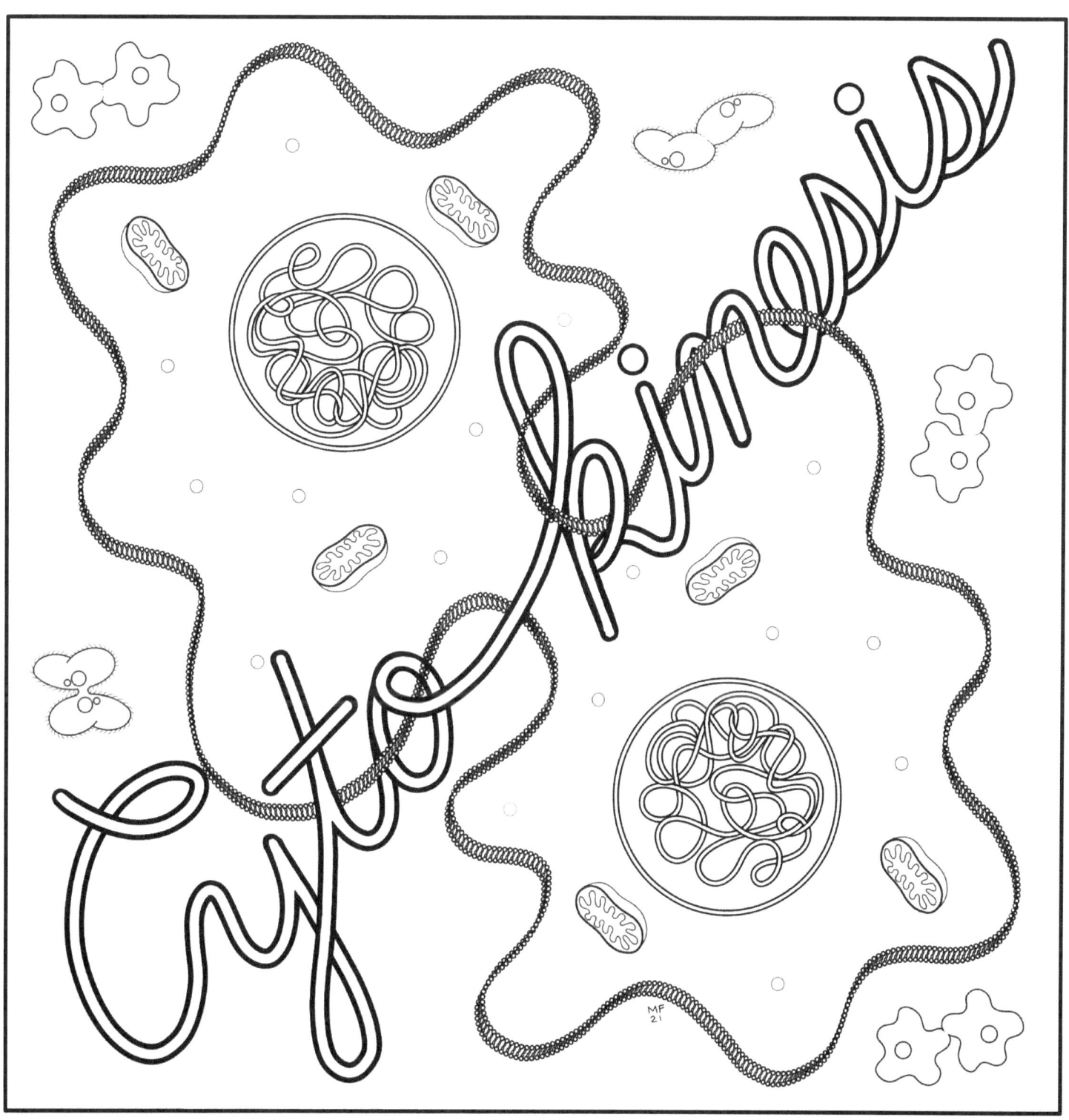

Definition: _____
_____
_____
_____
_____

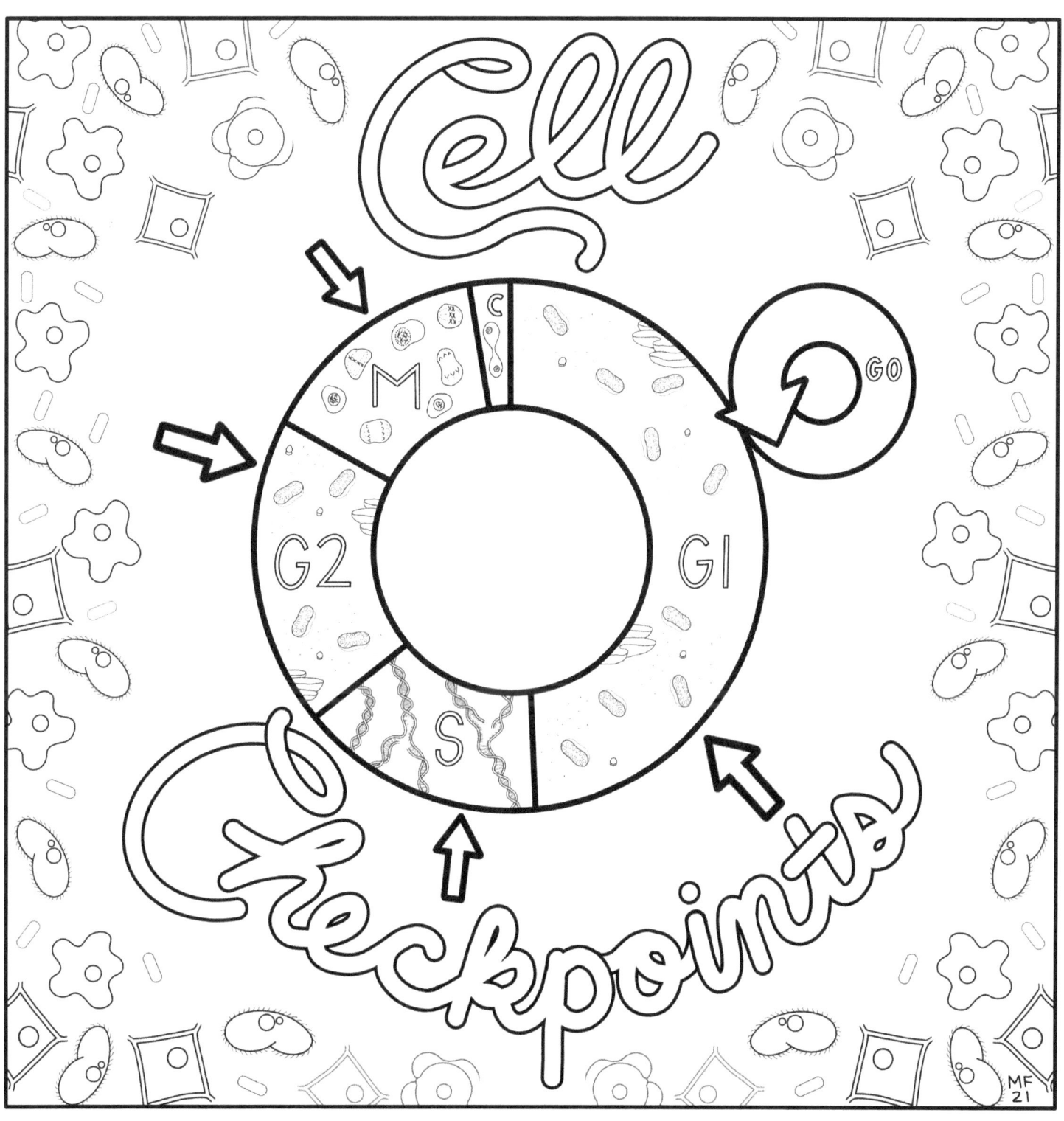

Definition: _____
_____
_____
_____
_____

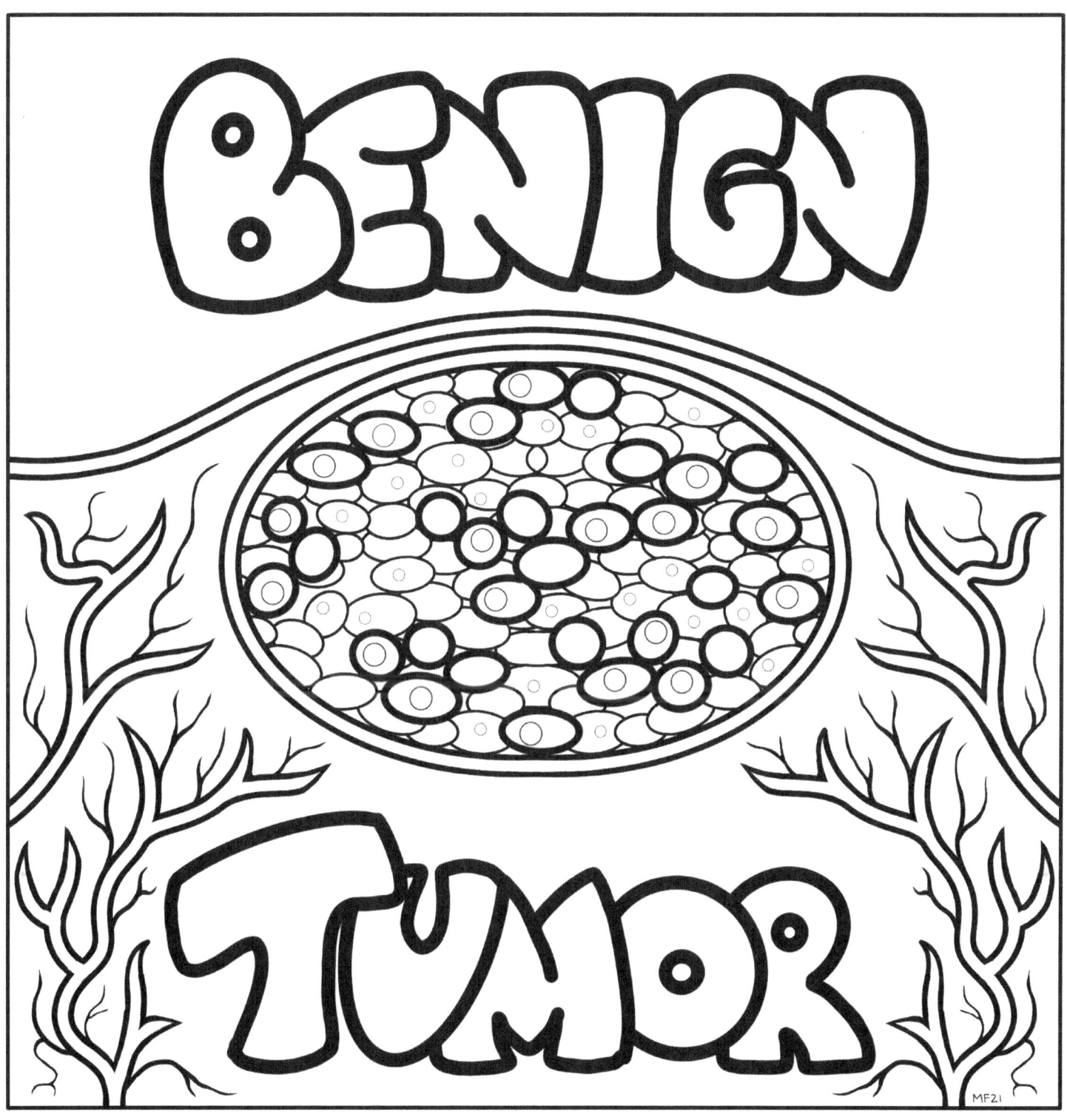

Definition: _____
_____
_____
_____
_____

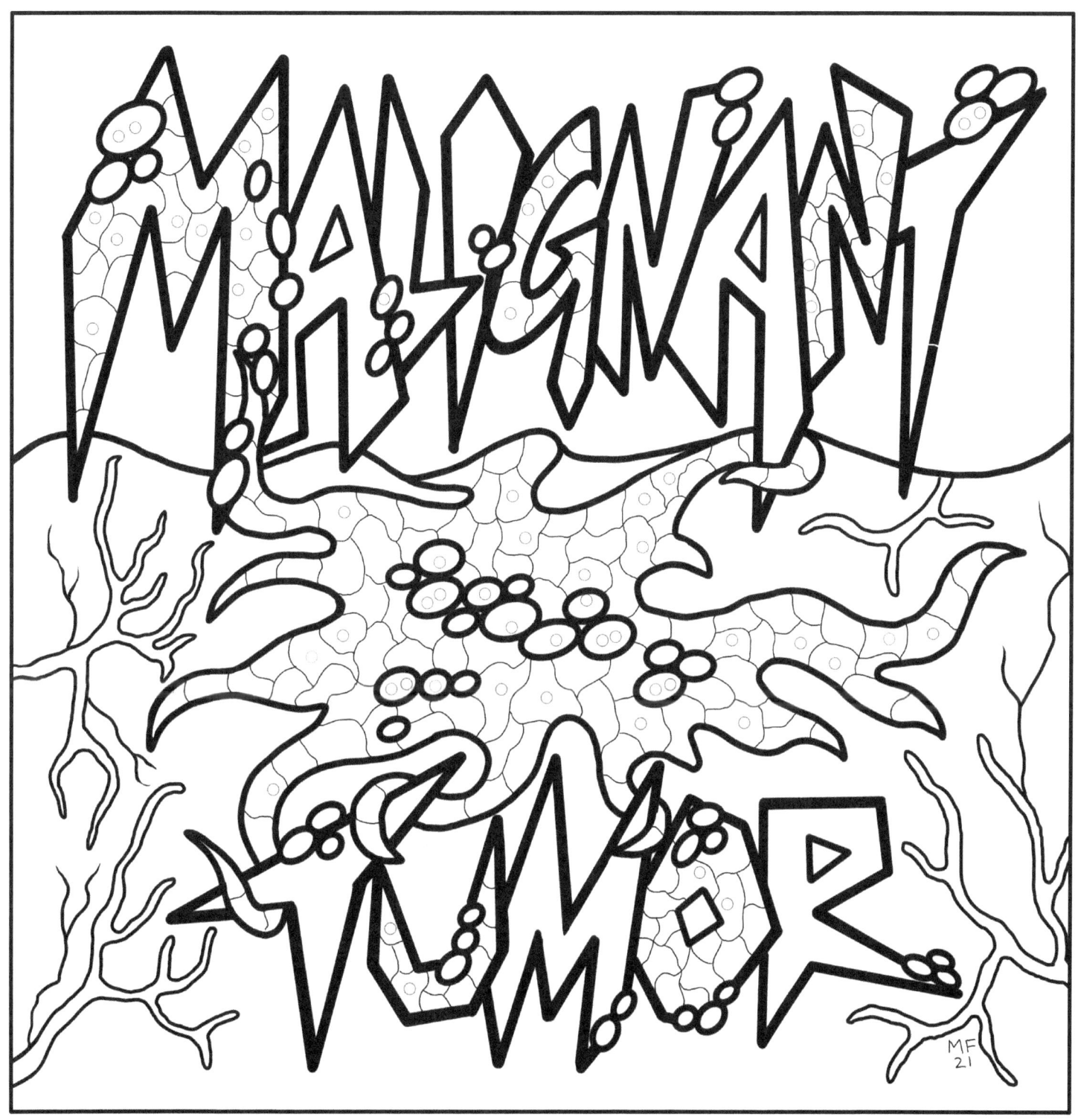

Definition: _____
_____
_____
_____
_____

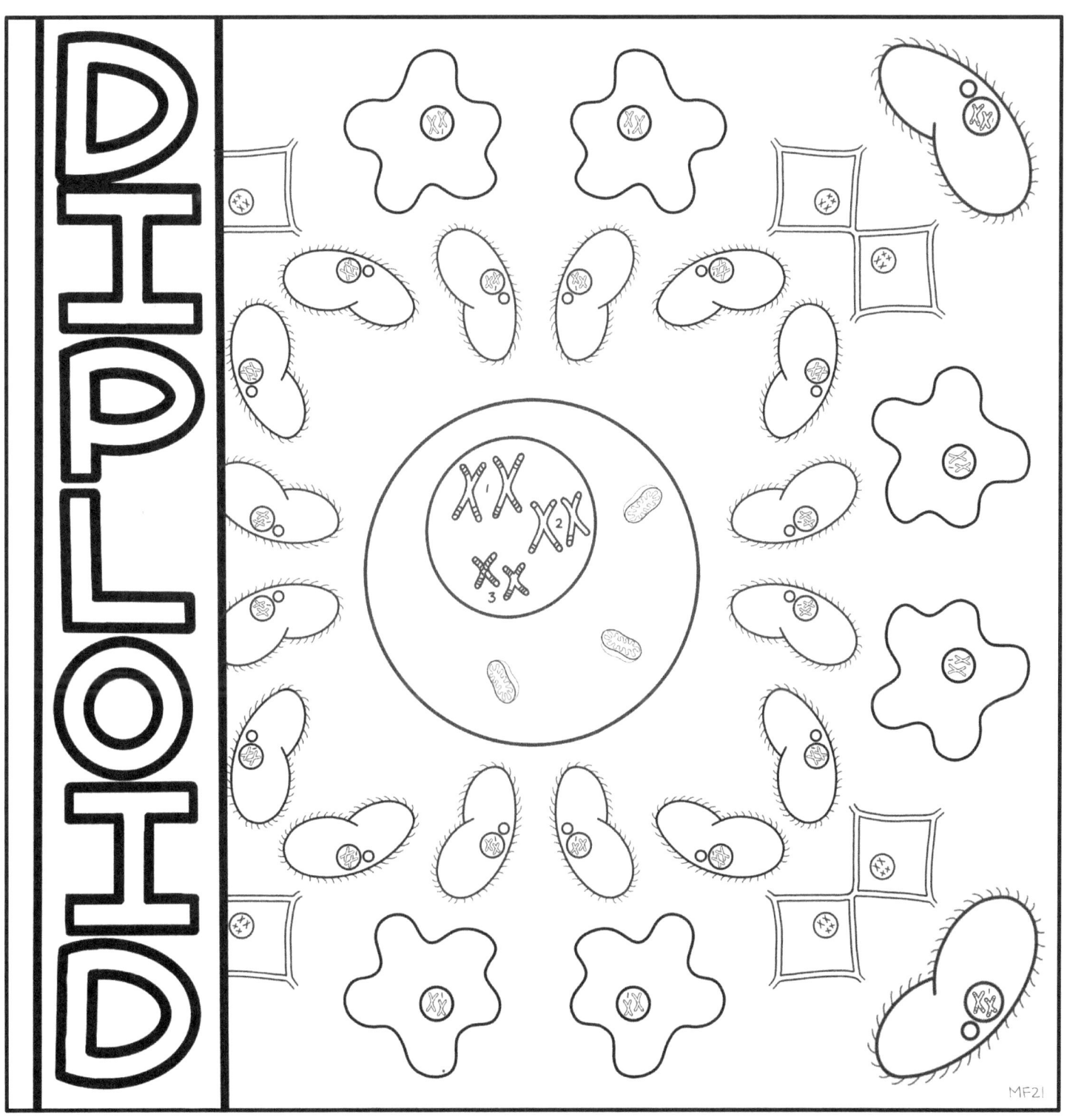

Definition: _____
_____
_____
_____
_____

Endoplasmic reticulum: A network of membrane-bound tubules and sacs that extend throughout the cytoplasm of eukaryotic cells. There are two types of ER, rough endoplasmic reticulum (RER) which has ribosomes attached and smooth endoplasmic reticulum (SER) which does not have ribosomes attached. RER is involved with the synthesis of phospholipids and processing of proteins. SER is involved with lipid metabolism and synthesis. SER also stores calcium ions.

Eukaryote: Complex cells which contain a membrane bound nucleus and membrane bound organelles. Eukaryotes are complex cells because they have a wide variety of organelles and can be classified into four kingdoms: Protist, Plant, Animal, and Fungus. "Eu" means true, and "Karyon" means kernel which is how the nucleus was described when the cells were viewed under a microscope.

Fungus cell: A specific eukaryotic cell that is the basic structural and functional unit of organisms classified in the Fungus Kingdom. Fungus cells, like all eukaryotic cells, are characterized by having membrane-bound organelles and a distinct nucleus containing the cell's genetic material. Fungus cells have cell walls made of chitin as their outermost covering. There are no chloroplasts in fungus cells because they are heterotrophs. Fungus cells have endoplasmic reticulum, acidic vesicles, peroxisomes, ribosomes, mitochondria, cytoplasm, and Golgi apparatus.

Golgi apparatus: Membrane-bound stack of cisternae found in eukaryotes which process, sort, and package proteins and lipids that are made within the cell. The Golgi apparatus transports lipids and proteins out of the cell through exocytosis.

Interphase: The longest phase of the cell cycle, where the cell spends most of its life cycle. During interphase, a cell grows, develops, reproduces its DNA, and prepares for cell division. Interphase has three stages: Gap 1 (G1), Synthesis (S), and Gap 2 (G2). G1 is where the cell grows and develops. S is where the cell duplicated its genome. G2 is where the cell prepares for division if it is not in G0. G0 is where cells remain if they are no longer dividing.

Lysosome: Membrane-bound organelle found in eukaryotes such as animal and protist cells. Lysosomes contain enzymes which break down and recycle macromolecules, cellular waste, food, and pathogens.

Malignant tumor: Abnormal mass of cells which divide and grow too quickly due to mutations in the DNA. Malignant tumors metastasize, which means they spread into surrounding tissues and can spread into new regions of the organism's body.

Metaphase: A stage within mitosis and meiosis where the chromosomes line up in the center of the cell before migrating to opposite poles of the cell.

Microtubules & Microfilaments: Cytoskeleton filaments found in eukaryotic cells which play an integral part in maintaining the cell shape, helping with cell motility, and intracellular transport.

Mitochondria: Membrane-bound organelle found in eukaryotes which is the site for ATP production during cellular respiration (aerobic respiration). ATP is the source of energy for cells, which is why mitochondria are often referred to as the powerhouse of the cell. Mitochondria used to be free living prokaryotic cells and have their own cDNA and ribosomes. Millions of years ago mitochondria formed an endosymbiotic relationship with eukaryotic cells.

Mitosis: Eukaryotic cellular division which results in two genetically identical daughter cells that each have the same number of chromosomes as the parent cell. Mitosis is the process by which organisms grow, develop, make repairs, and asexually reproduce.

Nucleus: Membrane-bound organelle found in eukaryotes which houses the cell's genetic information as chromosomes. The nucleus is the site for DNA synthesis and DNA transcription into mRNA. The nucleus is also where the nucleolus is housed. The nucleolus produces ribosomes.

Organelle: A specialized compartment within cells which has a specific structure and function. Most organelles are membrane-bound except for ribosomes. Organelles perform specific activities to aid in growth, functioning, survival, and defense of cells.

Peroxisome: Membrane-bound organelles found in eukaryotic cells which breakdown lipids using enzymes such as catalase to break down hydrogen peroxide into water and oxygen. Peroxisomes also break down nucleotides.

Phospholipid: Biomolecule composed of two hydrophobic fatty acid chains and a hydrophilic head made up of glycerol and phosphate. Phospholipids create a semi-permeable bilayer that is the main component of cell membranes (plasma membranes).

Plant cell: A specific eukaryotic cell that is the basic structural and functional unit of organisms classified in the Plant Kingdom. Plant cells, like all eukaryotic cells, are characterized by having membrane-bound organelles and a distinct nucleus containing the cell's genetic material. Plant cells' most unique features are a cell wall made of cellulose, chloroplasts, and a central vacuole. Plant cells have endoplasmic reticulum, acidic vesicles, peroxisomes, ribosomes, mitochondria, cytoplasm, and Golgi apparatus. Specialized plant cells may have amyloplasts or chromoplasts.

Prokaryote: Simple cells which do not contain a membrane bound nucleus and membrane bound organelles. Prokaryotes are bacteria cells and make up the Archaebacteria and Eubacteria kingdoms. "Pro" means before, and "Karyon" means kernel which symbolizes the nucleus. Prokaryotes are the first cells to evolve, the eukaryotes followed later.

Prophase: The first stage of mitosis and meiosis where the nuclear envelope dissolved while the chromatin condenses into chromosomes.

Protist cell: A specific eukaryotic cell that is the basic structural and functional unit of organisms classified in the Protist Kingdom. Protists can be single cellular or multicellular. Protist cells, like all eukaryotic cells, are characterized by having membrane-bound organelles and a distinct nucleus containing the cell's genetic material. Protist cells have endoplasmic reticulum, peroxisomes, ribosomes, mitochondria, cytoplasm, and Golgi apparatus. Some protists have a macronucleus and a micronucleus. Others have multiple nuclei. Aquatic protists have contractile vacuoles to regulate water flowing into the single cell, while some also have a red eye spot to help detect light. Protist cells have a large range of unique organelles depending on whether they are plant-like, animal-like, or fungus-like protists.

Ribosome: Nonmembrane-bound organelle found in both prokaryotes and eukaryotes which synthesize protein.

Telophase: The last stage of mitosis which precedes cytokinesis. During telophase, the chromosomes who have migrated into two groups at opposite poles of the cell will begin to decondense while new nuclear envelopes form around the genetic material. At the end there will be two new nuclei which will be divided into two cells by cytokinesis.

Vacuole: Membrane-bound organelles which store water and other important nutrients and biomolecules in eukaryotic cells. The large central vacuole found in plant cells maintain turgor pressure by storing water which gives the plant support to stand upright and maintain healthy firm stems and leaves.

Vesicle: Small temporary membrane-bound organelles found in eukaryotic cells which are used for transport, digestion, storage, and communication.

www.ingramcontent.com/pod-product-compliance
Lightning Source LLC
Chambersburg PA
CBHW062334220526
45469CB00008B/2715